Localization, Mapping and SLAM in Marine and Underwater Environments

Localization, Mapping and SLAM in Marine and Underwater Environments

Editors

Antoni Burguera
Francisco Bonin-Font

MDPI • Basel • Beijing • Wuhan • Barcelona • Belgrade • Manchester • Tokyo • Cluj • Tianjin

Editors

Antoni Burguera	Francisco Bonin-Font
Universitat de les Illes Balears	Universitat de les Illes Balears
Spain	Spain

Editorial Office
MDPI
St. Alban-Anlage 66
4052 Basel, Switzerland

This is a reprint of articles from the Special Issue published online in the open access journal *Journal of Marine Science and Engineering* (ISSN 2077-1312) (available at: https://www.mdpi.com/journal/jmse/special_issues/localization_mapping_SLAM).

For citation purposes, cite each article independently as indicated on the article page online and as indicated below:

LastName, A.A.; LastName, B.B.; LastName, C.C. Article Title. *Journal Name* **Year**, *Volume Number*, Page Range.

ISBN 978-3-0365-5497-6 (Hbk)
ISBN 978-3-0365-5498-3 (PDF)

Contents

About the Editors

Antoni Burguera

Antoni Burguera is an associate professor at the Departament de Matemàtiques i Informàtica (Universitat de les Illes Balears, Spain), where he teaches different subjects in Computer and Industrial Engineering.

He is part of the Systems, Robotics and Vision group (SRV), and is engaged in research tasks on mobile robotics, especially underwater robotics. His research interests focus on localization and SLAM, both using visual and acoustic sensors, as well as on Artificial Intelligence and Deep Learning. A list of his main research contributions is available at http://srv.uib.es/antoni-burguera-burguera/.

Francisco Bonin-Font

Francisco Bonin-Font is an associate professor at the Departament de Matemàtiques i Informàtica (Universitat de les Illes Balears, Spain), where he teaches different subjects in Computer and Industrial Engineering.

He is part of the Systems, Robotics and Vision group (SRV), and is engaged in research tasks on mobile robotics, especially underwater robotics. His research interests focus on mobile robot localization and SLAM and computer vision. A list of his main research contributions is available at http://srv.uib.es/francisco-jesus-bonin-font/.

Preface to "Localization, Mapping and SLAM in Marine and Underwater Environments"

The use of robots for field applications in underwater and marine environments is rapidly growing. These applications have one common requirement: to properly model the environment and estimate the robots' poses. Even though several mapping, SLAM, and target detection and localization methods exist, marine and underwater environments have some particularities that need to be addressed, such as reduced vision range, water currents, communication problems, sonar inaccuracies and unstructured environments.

This Special Issue aims to highlight the current research trends related to the topics of underwater localization, mapping and SLAM as well as target detection and localization. To this end, it presents seven papers from leading scholars in the field and demonstrates the diversity of approaches and methods that are nowadays being explored to improve the performance of underwater robots.

Antoni Burguera and Francisco Bonin-Font
Editors

Journal of
Marine Science and Engineering

MDPI

Article

ROV Navigation in a Fish Cage with Laser-Camera Triangulation

Magnus Bjerkeng [1,*], Trine Kirkhus [1], Walter Caharija [2], Jens T. Thielemann [1], Herman B. Amundsen [2], Sveinung Johan Ohrem [2] and Esten Ingar Grøtli [1]

1 SINTEF Digital, 0373 Oslo, Norway; trine.kirkhus@sintef.no (T.K.); jens.t.thielemann@sintef.no (J.T.T.); esteningar.grotli@sintef.no (E.I.G.)
2 SINTEF Ocean, 7010 Trondheim, Norway; walter.caharija@sintef.no (W.C.); herman.biorn.amundsen@sintef.no (H.B.A.); Sveinung.ohrem@sintef.no (S.J.O.)
* Correspondence: magnus.bjerkeng@sintef.no

Abstract: Aquaculture net cage inspection and maintenance is a central issue in fish farming. Inspection using autonomous underwater vehicles is a promising solution. This paper proposes laser-camera triangulation for pose estimation to enable autonomous net following for an autonomous vehicle. The laser triangulation 3D data is experimentally compared to a doppler velocity log (DVL) in an active fish farm. We show that our system is comparable in performance to a DVL for distance and angular pose measurements. Laser triangulation is promising as a short distance ranging sensor for autonomous vehicles at a low cost compared to acoustic sensors.

Keywords: autonomous navigation; range sensing; inspection and maintenance; 3D vision

Citation: Bjerkeng, M.; Kirkhus, T.; Caharija, W.; Thielemann, J.T.; Amundsen, H.B.; Johan Ohrem, S.; Ingar Grøtli, E. ROV Navigation in a Fish Cage with Laser-Camera Triangulation. *J. Mar. Sci. Eng.* **2021**, *9*, 79. https://doi.org/10.3390/jmse9010079

Received: 8 December 2020
Accepted: 11 January 2021
Published: 13 January 2021

1. Introduction

Nearly half of the earth's land is used for food production, and marine resources can help feed the growing population. The number of fish farms is increasing rapidly [1]. Typically, the fish is raised in open sea net cages, which consist of a floating collar, a net pen and a mooring system. These cages are in natural marine environments, and fish that escape from these environments may cause harm to the environment and its related food chain. To minimize the escape caused by the failure of the net cage, the net must be inspected routinely [2]. Net inspections today are commonly performed either by divers or by human-piloted remotely operated vehicles (ROVs). The ROV operations are challenging for the pilot as they require both precise maneuvering and a keen eye for detail in order to detect failures in the net cage from the video stream.

One of the main problems in applying autonomous underwater vehicle to fish cage inspection is the automatic detection and tracking of net pens. This is to maintain a safe distance from the net pen and to ensure complete coverage of the cage during the inspection. Cost constraints are in addition tight since the autonomous vehicle needs to be similar in cost, or ideally cost less than the human divers used today. Even though there are no industrial deployment of autonomous net inspection systems that we are aware of, it is an active topic in research [3]. There are also similar research programs within subsea oil and gas, [4], and seabed mapping [5]. Aquaculture net pens are especially challenging to inspect because their shape changes with the water current and due to biofouling changing their visual appearance [6].

Successful operation of autonomous underwater vehicles requires the ability to navigate, and to understand dynamic environments. There are many mature positioning systems which can position underwater vehicles. The long baseline method (LBL) and the ultrashort baseline method (USBL) use acoustic ranging relative to fixed beacons. These methods require pre-deployed and localized infrastructure, hence increasing the cost and the complexity of the operations [7,8]. Furthermore, underwater ranging systems are challenged by infrastructures prohibiting line-of-sight as, e.g., aquaculture sites [9]. Position measurements can be integrated with velocity measurements provided by an acoustic doppler velocity log (DVL) and an onboard digital compass. [10].

Several optical systems are used to get 3D data from underwater. Video cameras in combination with markers are commonly used for autonomy and navigation [11], for underwater stereo [12,13] or photogrammetry [14], which will need unique, non-repetitive features in the scene to estimate the disparity and thereby the depth measurement. Due to the repetitive structure of the nets to be inspected, stereovision is not well suited in our use-case.

Structured Light method uses typically a DMD (digital mirror device) to project a single pattern or a series of spatially coded patterns to get highly accurate 3D measurements of the scene in real-time. In [15], Multi-Frequency Phase Stepping patterns are used to acquire high resolution 3D data from a static scene in turbid water.

Scanning LIDARs (Light Detection and Ranging) are used for inspection tasks underwater [5]. Due to the scanning nature and the capturing time of this method, it needs a compensation for the relative motion of the vehicle. Flash-LIDARs do not include a scanning and provide real-time 3D data with depth precision below 1 cm at high signal levels, at 10 Hz [16].

Laser triangulation systems typically project a laser line [17–21] or point [22] onto the scene to triangulate distances between the laser and a camera. Commercial systems are also available, e.g., from 2GRobotics (www.2grobotics.com). These systems will need a scanning device for getting 3D data from the whole scene, which makes them slow, mechanically complex and expensive. We needed a cheap system, and to reduce complexity we wanted to use the ROV's built-in camera. The suggested solution uses two laser lines to enable detection of a plane from one single image of the projected lasers. Parallel lines were chosen to get an optimal baseline geometry between the camera and both the laser line sources—this also results in a compact system suited for mounting on the available ROV, and also enables estimation of both pitch and yaw.

The algorithms which interpret the data from these sensors, to achieve autonomy, were first addressed in a probabilistic framework by [23], which is known as the Simultaneous Localization and Mapping (SLAM) problem. In view-based or dense SLAM, visual odometry is performed by comparing two complete views [24], e.g., by registering overlapping perceptual data, for example, optical imagery [25] or sonar bathymetry [26]. Unstructured underwater environments pose a more challenging task for feature extraction and data association than terrestrial environments. Hence, the application of feature-based SLAM frameworks has so far had limited success in real-world underwater environments [27].

To enable cost effective underwater SLAM for net inspection, this paper proposes using laser-camera triangulation consisting of two laser lines and one camera for pose estimation from one image. By assuming the net wall can be approximated to a plane, using two laser lines enables fitting a plane to the net cage's wall based on one image only. This enables estimation of pose relative to the net pen in real time. The partial pose of a camera with respect to an observed net pen can be used for closed-loop net-following control.

The on-board camera used for net pen inspection was used for the laser triangulation. The only extra hardware needed are two lasers and their power supply, which drives down cost compared to acoustic sensors. No synchronization circuit or communication between laser and camera is needed as we run the lasers continuously.

2. Materials and Methods

2.1. The ROV and the Sensors Employed

The ROV employed in the trials is an Argus Mini, manufactured by Argus Remote Systems AS, shown in Figure 1. It is an observation class ROV specifically built for inspection and intervention operations in shallow waters, and meant to serve scientific purposes as well as the offshore, inshore, and fish farming industries. The Mini weights

90 kg with dimensions L × B × H = 0.9 m × 0.65 m × 0.6 m and is designed around six ARS800 thrusters. Four of the thrusters are placed in the horizontal plane, while the other two are placed in the vertical plane, hence guaranteeing actuation in 4 degrees of freedom (DOFs), i.e., surge, sway, heave, and yaw. The ROV is passively stabilized by gravity in roll and pitch.

Figure 1. The Argus Mini remotely operated vehicle (ROV), courtesy of Argus Remote Systems AS.

The Argus Mini is equipped with 5 sensors: a SONY FCB-EV7100 Full HD camera, a fluxgate compass, a depth sensor, a gyro, and a Nortek DVL 1000 velocity sensor. In addition, position measurement is provided by a Sonardyne USBL system that consists of a Micro-Ranger Transceiver mounted onboard the support vessel and a Nano Transponder mounted on the vehicle. The ROV contains no sensors for direct measurement of acceleration.

The Nortek DVL is forward-looking, i.e., the instrument is mounted on the front of the ROV, pointing in the x-direction of the body/vehicle frame. This unconventional DVL configuration is employed with the purpose of enabling DVL lock on submerged vertical structures present in the aquaculture context, such as net cages of large fish farming cages (50 m in diameter). Such features of the DVL instrument, combined with its ranging features, are utilized in [28] to estimate the ROV distance and heading relative to a net cage and validate a guidance law for autonomous net following.

2.2. The ROV Control System

The company SINTEF employs its in-house control systems on the Argus Mini ROV. The ROV has three operational modes: manual (assisted with auto-heading (AH) and auto-depth (AD)), dynamic positioning and net-following. Relevant to the experiment presented here is the net-following controller, which uses feedback from the forward looking DVL. Net-following makes the ROV autonomously traverse aquaculture net cages at a given depth. The method exploits the four range measurements provided by the DVL beams to approximate the geometry of the net cage in front of the ROV as a plane through a least-squares regression. It then calculates the ROV position and orientation relative to this local plane. The relative position and orientation are subsequently fed as inputs to a nonlinear line-of-sight guidance law [29]. Further details on the employed net following (NF) guidance as well as the net cage geometry approximation by use of DVL range measurements can be found in [28].

A 4 DOF extended Kalman filter is also running to assist the dynamic positioning. The Kalman filter fuses the position measurements provided by the USBL, the velocity measurements form the DVL and the mathematical model of the ROV to estimate the vehicle state [30].

2.3. Sea Trials for Data Collection

The trials were executed at the SINTEF ACE Tristeinen aquaculture facility shown in Figure 2. SINTEF ACE is a full-scale laboratory designed to develop and test new aquaculture technologies under realistic conditions. The tests were performed inside a cage for salmon farming, in full operational state. ROV operations in fish farms are commonly performed inside the fish cages, not outside. This is due to the presence of ropes, chains and mooring lines on the outside of the cages which the ROV's tether can get tangled into. When operating inside the cage, the fish will sometimes obstruct the cameras and sensor measurements, but the severity of this compared to the ROV being tangled is low. The cage has a cylindrical shape with a conical bottom, where the upper diameter is approximately 50 m and the total depth is about 30 m. The cage used in the trial is equipped with double nets in the regions around the main ropes to secure these regions against fish escapes. This double net setup is used in all fish cages operated by the company operating at SINTEF ACE, but it is unknown if this is standard for other companies. As will be shown, the presence of the double nets influences the quality of the distance measurements when using the DVL. The nets at the Tristeinen facility are square, with a mesh width of 33 mm. They had been cleaned eight days prior to the trials. The nets originally had a green coating, but some of this seems to have worn off. Fish population was approximately 190.000 individuals, which is normal.

Figure 2. SINTEF ACE, a full-scale laboratory facility designed to develop and test new aquaculture technologies.

The tests consisted of pointing the laser lines against the cage net and simultaneously recording HD videos at 60 frames per second (fps), while having the ROV executing several net cage traverses at constant depth by utilizing the NF guidance, interrupted by short intervals where the ROV was placed in dynamic positioning (DP) mode. Such a configuration allows the direct comparison of the laser-camera system capabilities with the DVL capabilities during the execution of net following tasks, which is highly relevant in the context of subsea aquaculture operations [9].

2.4. D Data Camera–Laser Line Triangulation

To get 3D data from the fish cage net, we chose to use the method of triangulating between two laser lines and one camera. Due to the repetitiveness of the pattern in the cage's net, we chose a method not relying on correlating features in the scene and rather projecting the pattern (here, two laser lines). This also enables 3D data in the dark, e.g., at night or at larger depths. A blue laser was chosen to limit the effect of light scattering particles and attenuation of light in the water.

Two laser lines (OdicForce Lasers's 80 mW Blue, 450 nm, Adjustable Locking Focus Direct Diode Module Line Pattern) where chosen to enable estimation of both the distance and the position relative to the net wall—assuming the wall is planar. The camera used was the camera available on the ROV (SONY FCB-EV7100 Full HD). The actual setup is shown in Figure 3.

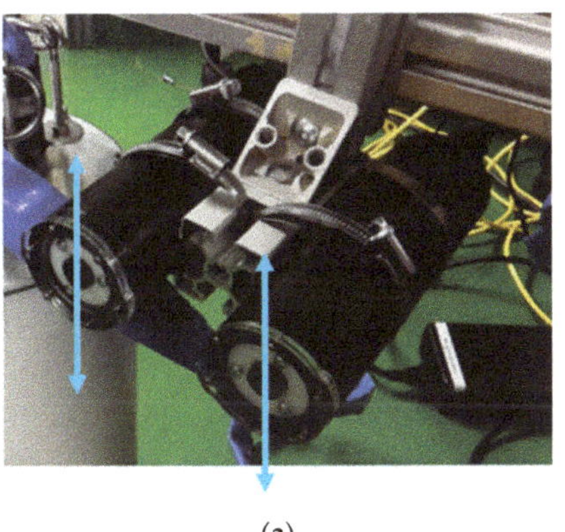

(**a**)

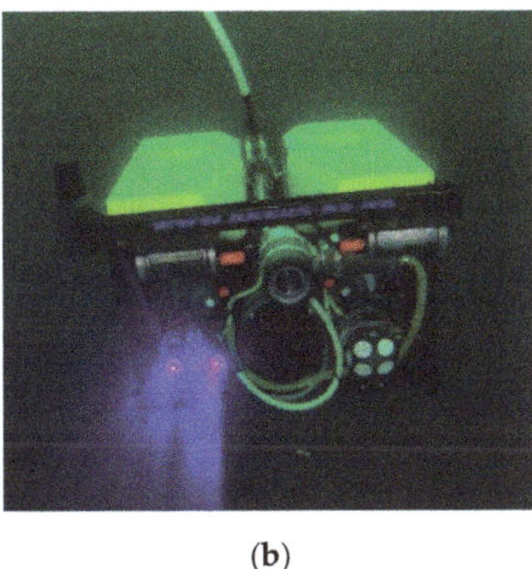

(**b**)

Figure 3. (**a**) The laser lines are mounted inside watertight tubes from Blue Robotics. The laser lines are perpendicular to the plane containing the two tubes and the camera. (**b**) The laser lines are mounted onto the ROV. The camera used is the on-board ROV camera, seen in the center of the ROV.

We calibrated the system by capturing images of submerged checkerboards where the laser lines also were projected onto the checkerboard (Figure 4). The calibration was performed underwater. Then, the refraction glass–water is seen as a lens effect handled by the calibration routines; this effect is also reduced by using a dome shaped glass to ensure perpendicular surface from camera lens to glass. Attenuation and scattering due to the water and its turbidity is handled by using enough light matching the distances we operate. This aligns with the conclusion in [31]. The calibration was performed by moving the ROV to get a good dataset with different viewing angles. Using Zhang's method of camera calibration [32], we recovered camera distortion parameters using openCV's camera model implementation of the camera matrix A, and the distortion coefficients K,

$$A = \begin{bmatrix} c_x & 0 & c_x \\ 0 & f_y & c \\ 0 & 0 & 1 \end{bmatrix}, \tag{1}$$

where f_x, f_x are the focal lengths and c_x, c_y are the principal point; and the distortion parameters,

$$K = [k_1, k_2, k_3, p_1, p_2], \tag{2}$$

where k_1, k_2, k_3 are the radial distortion parameters and p_1, p_2 are the tangential distortion parameters.

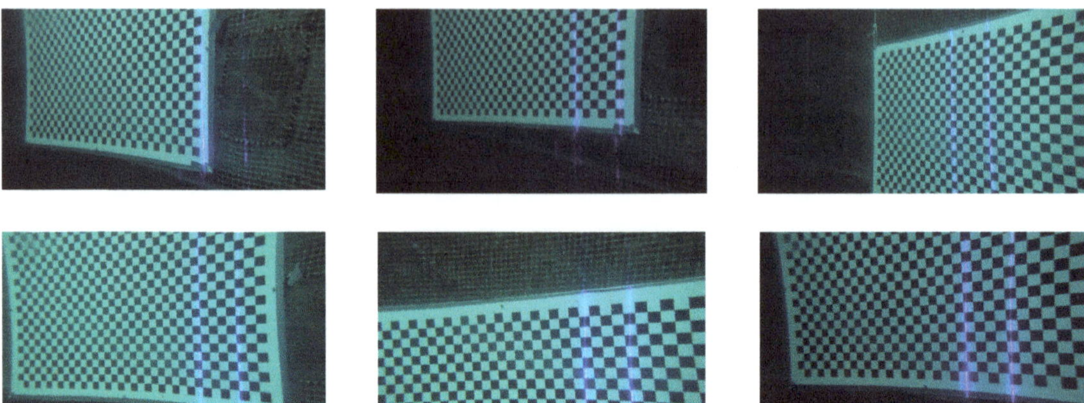

Figure 4. Example of calibration images for the camera-laser 3D measurement. The laser lines can be seen in the left part of the checker board.

We also recover the 3D position of the laser lines on the checkerboards. This information was used to recover the plane parameters for the two projected laser planes, meaning that we could perform laser triangulation by intersecting camera rays (lines) with the laser plane for *xyz* recovery. Assuming camera as the origin, all camera rays can be expressed as points

$$p = d \cdot L \tag{3}$$

where $L = [x_L, y_L, z_L]$ is the direction of the ray and d is the position along the ray. The laser plane can be defined as

$$(p - p_0) \cdot n = 0, \tag{4}$$

where p_0 is a point on the plane, n is the normal to the plane and p are the points of the plane. To determine d we solve for

$$d = \frac{(p_0 - l_0) \cdot n}{l \cdot n}, \tag{5}$$

meaning that we can find the *xyz* point of the intersection as $d \cdot L$.

To get the distance, yaw and pitch between the ROV and net cage wall, the laser lines were detected in images of the net when the two laser lines were projected. The laser line points' positions in the images were located with sub pixel accuracy and looked up in the calibration lookup table to get the points' absolute x, y and z distance from the camera's center. The resulting positions from both laser lines were by fitted to a plane using MLESAC (Maximum Likelihood Estimation SAmple. Consensus) [33]. MLESAC is a generalization of the RANSAC (RANdom SAmple Consensus) algorithm picking a subset of points, fitting a plane and searching for the plane with highest maximum likelihood to all points. MLESAC also brings robustness improvements relative to the original RANSAC algorithm. From the plane parameter, we get the distance to the plane, the yaw and pitch angles of the camera relative to the net wall. Example images and corresponding fitted planes are in Figure 5. We handle the detected laser lines as point sets (search for points per image row) and do not try to make them into lines. This makes us robust for outlier detections and the fact that the nets have holes which reduces the number of line points detected. Points from both laser lines are used simultaneously in the MLESAC algorithm.

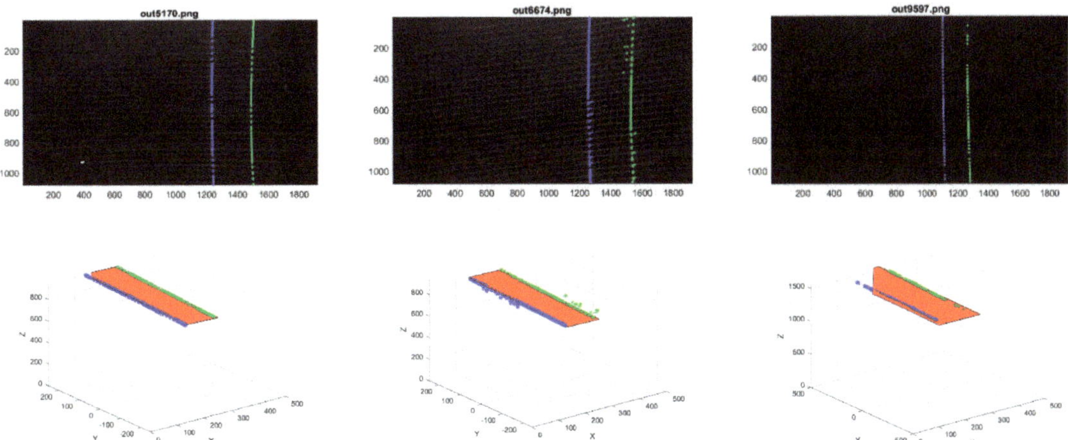

Figure 5. Example of ROV's laser line images for the camera–laser 3D measurement and the fitted plane representing the net cage wall. Upper row is the images, and lower row the plane fitted to the x,y,z positions of the points; units are in mm.

3. Results

This section compares the sensor readouts from a DVL and the laser-camera system for an experiment where an ROV is navigating inside a fish cage. Fish were swimming in the cage during the experiment, which contributes significantly to the noise. The DVL measurements are processed at the sensor and are filtered which can affect the apparent signal smoothness. The laser triangulation signal is the raw data, with no outlier rejection, smoothing, nor filtering applied. A graphic showing the geometry of the measurement setup is shown in Figure 6.

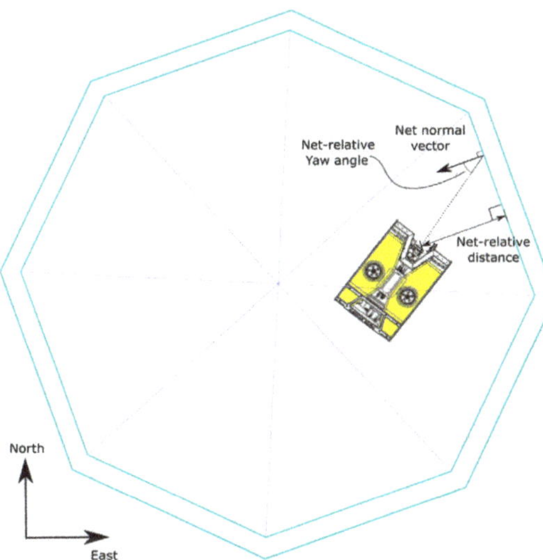

Figure 6. The geometry of the fish cage and the measured net distance and yaw angle in the horizontal North-East plane. The ROV is roll and pitch stable, but the pitch varies ±10 degrees around the zero point, so the horizontal assumption is not perfect. The yaw angle is calculated by projecting into the North-East plane.

An indirect sensor-to-sensor calibration was performed to compare the DVL measurements with the camera measurements. The camera position relative to the DVL position was determined during the installation on the ROV by measuring the position of the mounting points. No closed-loop extrinsic calibration was performed to precisely position the DVL with respect to the camera. The distance calibration consists of: (1) a manual time synchronization to shift the camera signal to be in step with the DVL time with a static bias; and (2) a single static bias of 0.37 m, added to the laser measurement to bring the camera plane in line with the DVL plane. The yaw calibration only synchronizes the time, since the two sensors were mounted to have the same orientation—and should not need any signal correction. The net-to-ROV pitch angle is also measured but is not reported since the ROV is pitch stable—and the signal is small.

3.1. DVL vs. Laser Triangulation Depth Data

Figure 7 compares raw output data from the DVL and the laser triangulation distance measurements. The two sensors are largely in agreement. An attempt was made to use the USBL localization system as a third sensor to establish ground truth, and determine which sensor principle was more accurate in absolute terms. However, the dynamic nature of the net cages made the USBL data not usable for this purpose. The laser triangulation measurements are noisier than the DVL distance. The high level of agreement between the two sensors is surprising, given the open-loop calibration only along the depth axis to position the sensors relative to each other.

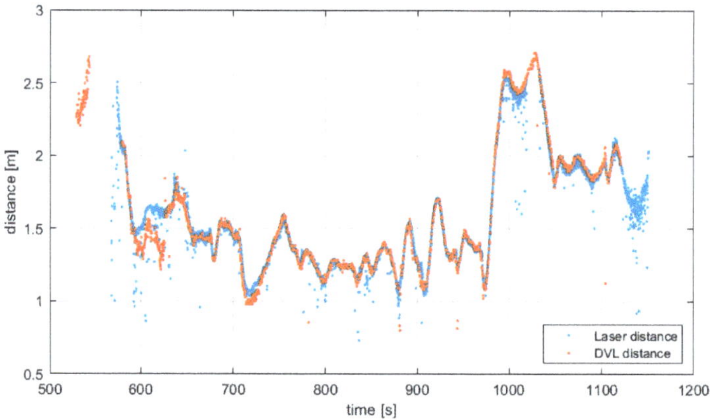

Figure 7. The doppler velocity log (DVL) distance measurement compared with the laser triangulation measurement.

There are two areas in the distance data we will look at in detail. The first is the disagreement around t = 650 s. The DVL and laser have a disagreement of around 25 cm. A picture from taken at that time is shown in Figure 8. It is seen that a double net is the cause of the problem. The laser line algorithm returns the distance to the closest net, as seen in Figure 9. We conclude that the laser triangulation distance is more reliable than the DVL for distance to the double net.

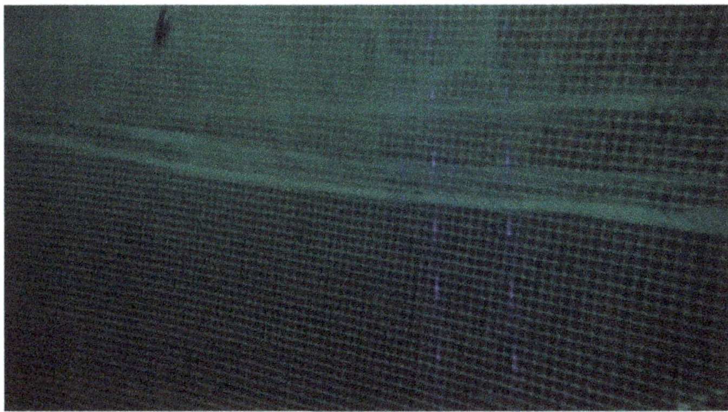

Figure 8. A picture taken at t = 650 s, where the triangulation distance disagrees with the DVL distance. The double net pen net in the upper part of the image confuses the DVL. Four laser lines are visible, two from each net.

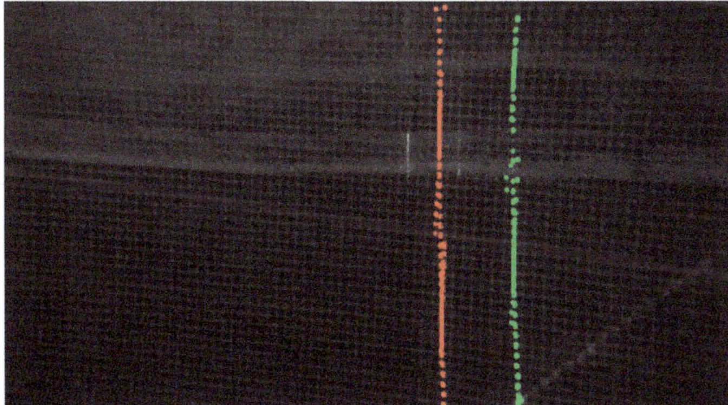

Figure 9. The laser line detection in the picture in Figure 10. The two brightest laser lines are found even when two extra less bright laser lines appear on a more distant net.

Figure 10. A picture taken at t = 1020 s, where the triangulation data is significantly more noisy than the DVL distance. Weak laser signal due to large distance (>2.5 m) to the net makes the laser triangulation loose data. The image is brightness-corrected for display purposes.

The next area of interest is at the maximum distance of 2.5 m, achieved at t = 1050. Here the laser triangulation is significantly noisier than the DVL. Looking at the picture at that time, seen in Figure 10, it is evident that low laser visibility due to turbid water is the problem, and the triangulation is close to the maximum range. For higher ranges than this—we would need a brighter laser. Introducing a band-pass filter, will filter out the stray light from other light sources and improve the laser line contrast; but on the other hand, it will filter out information needed for net pen inspection.

3.2. DVL vs. Laser Triangulation Yaw Data

The net-relative Yaw angles for both sensors are seen in Figure 11. The data shows that the ROV was looking at the net head-on with a deviation of 20 degrees in yaw, i.e., the net following controller is performing well. The loss of DVL signal towards the end of the dataset is due to the ROV ascent, which indicates that the laser triangulation may be more robust than the DVL in shallow waters, since the laser is less affected by reflection from the water surface than acoustic signals. The agreement between the two sensors is impressive given that no extrinsic calibration was performed apart from time synchronization. The increase in noise at the end of the dataset is due to the large amount of fish at that time.

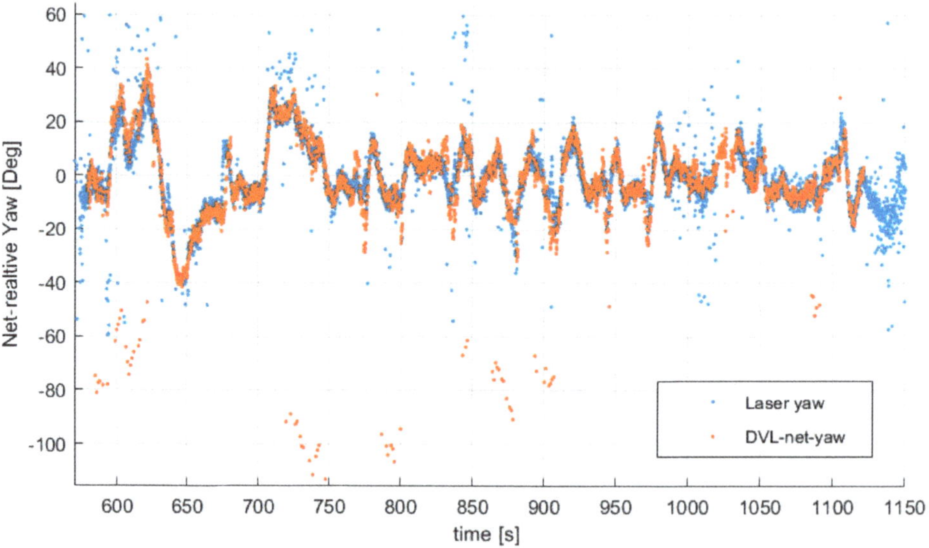

Figure 11. The DVL net-relative yaw compared with the laser triangulation measurement.

3.3. Kalman Filter Comparison

Figures 12 and 13 shows the trajectory traversed by the ROV during the experiment estimated by an extended Kalman filter. The curve seen is the circular fish cage. The main sensor driving the Kalman filter is a ship attached USBL system. Changes to the USBL either due to ship movement, net movement, or other error sources show up as jumps in the position estimates—showing that a USBL only system is not sufficient for robust net inspection. The dots are net positions relative to the ROV from the DVL and the laser triangulation. It is seen that the two distance sensors are mostly in agreement. It seems feasible to base a net-following controller on the output of the laser triangulation sensor. Figures 14–16 show the measured planes overlaid the Kalman pose estimate for only leg 3 to increase readability. The two sensors report similar data in this interval, showing that one could be exchanged for the other. The overlapping planes enable a well behaved pose-graph for a SLAM implementation.

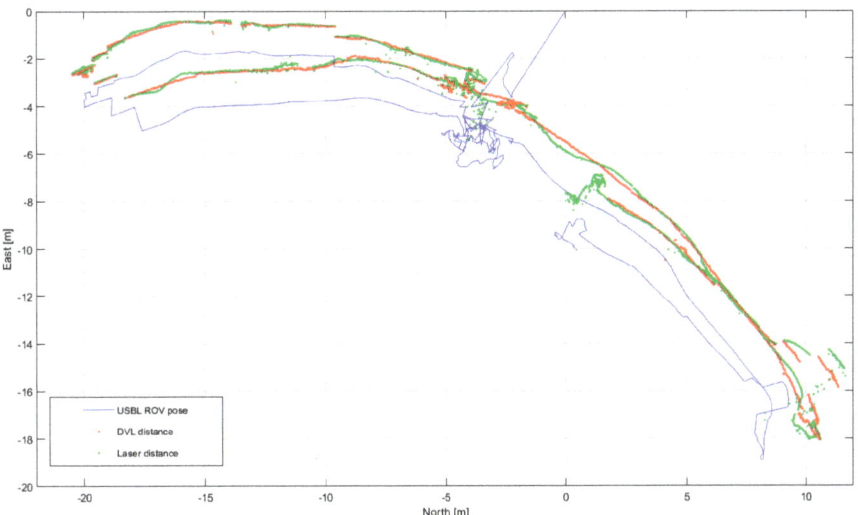

Figure 12. The position of the ROV is the blue line, the red dots are DVL distance measurements, and the green dots are laser measurements.

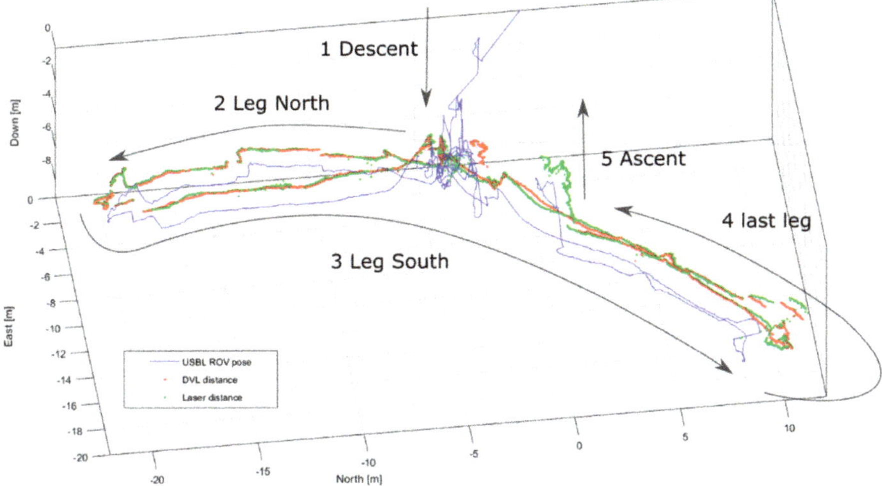

Figure 13. A Kalman filter estimate of the experiment showing the descent, navigation along the net pen, and the ascent.

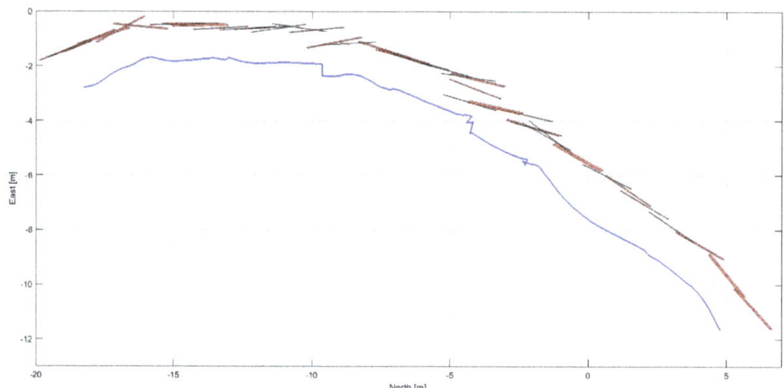

Figure 14. Kalman filter data from leg 3 of the experiment, with the DVL measured planes overlaid seen in the North-East plane.

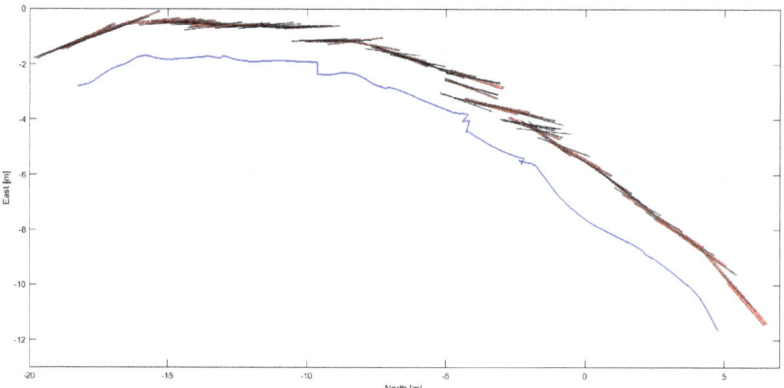

Figure 15. Kalman filter data from leg 3 of the experiment, with the laser measured planes overlaid seen in the North-East plane.

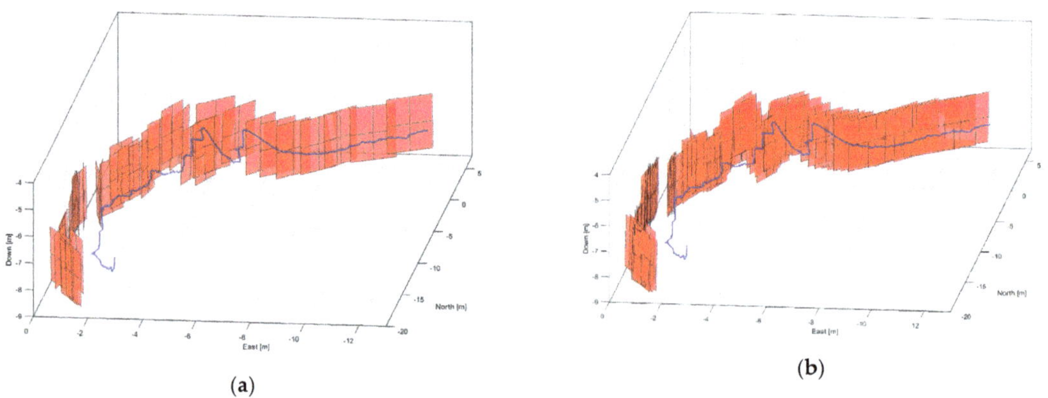

Figure 16. A 3d view of the measured planes. The DVL data is in (**a**), laser data in (**b**).

3.4. Quantization

This section compares a quantization effect seen on the DVL to the laser triangulation sensor. Figure 17 shows two sections of the distance measurements. A quantization error of around 1 cm sometimes affect the DVL data. The laser triangulation quantization error is less than mm scale. Any similar quantization issues were not seen on the DVL yaw data. It is not known if this issue is related to the acoustics, or is a sensor-specific issue.

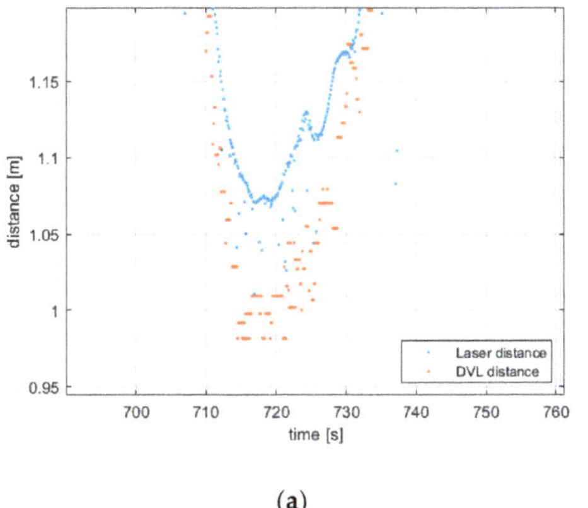

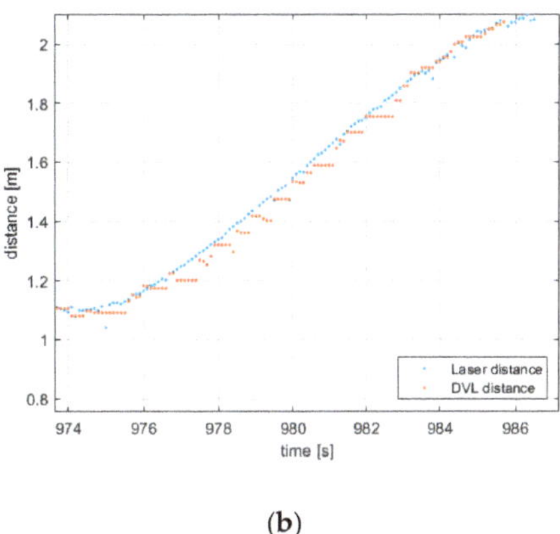

(**a**) (**b**)

Figure 17. Zoomed in distance measurements from the laser triangulation sensor and the DVL which show quantization effects.

3.5. Noise Comparison

This section attempts to compare the noise levels between the laser measurements and the DVL distance measurements. This comparison will not be strictly correct since there is no ground truth but may still be of interest. A linear Kalman filter was used to fuse the two distance measurements to obtain a quasi-ground-truth. The filter was manually tuned, and in addition, outliers were manually removed. The outliers were removed because they would dominate the result otherwise. We compared a time series where both sensors returned signals at the same time. The results are shown in Figure 18, with the overall result the DVL distance has a standard deviation measurement error of 2.9 cm, and the laser sensor has a comparable but slightly larger error of 3.2 cm.

This proof of concept shows that an autonomous vehicle with a camera can be cheaply upgraded with net cage sensing capacities. Most inspection vehicles have RGB cameras already installed, which is the expensive part of the sensor, and given that camera sensors improve in performance per dollar per year—this sensor type will continue to be attractive in the future. Cost wise, a DVL costs in the range of thousands to tens of thousands USD. In comparison—the two lasers and housings cost \$200, a factor of 25× to 100× in cost savings. These cost reductions are significant, especially for large fleets of inspection vehicles.

An added result is the verification that the DVL sensor measures the correct ROV-to-net distance. In, e.g., [9], a DVL is tested as a net cage navigation sensor, but since no independent measurement was available (it was not known that the DVL was unbiased for net pen measurements), comparative estimates indicate standard deviation of 3.2 cm for the laser system, and 2.9 cm for the DVL.

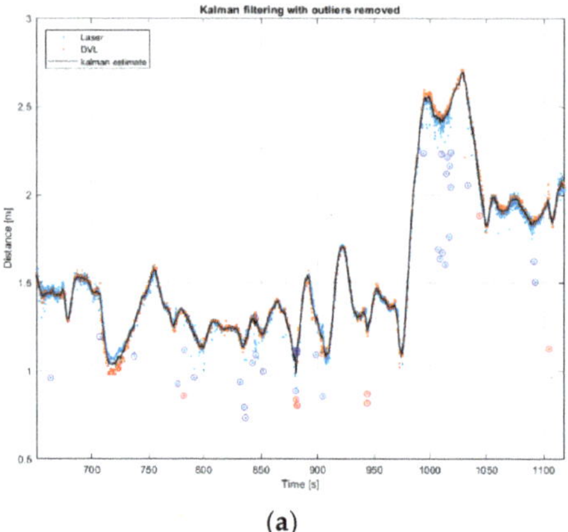

(a)

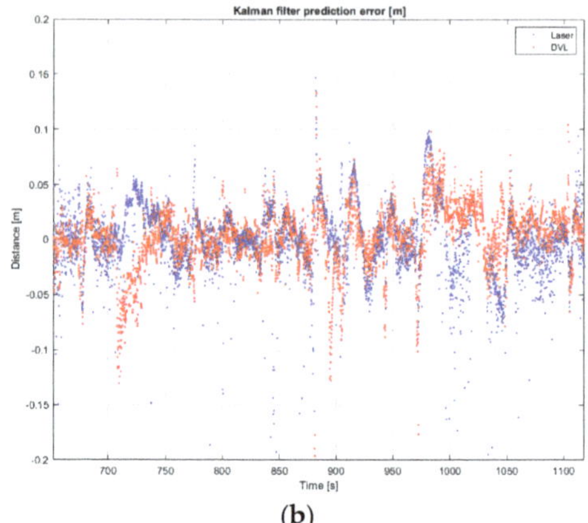

(b)

Figure 18. (**a**) shows a Kalman filter fusion of the distance measurements, with outliers removed. The outliers are indicated with circles. (**b**) Shows the Kalman prediction error which is the basis for estimating the noise levels for the two sensors.

One particular concern was that the double nets, an outer net and an inner net, resulted in systematic bias on the DVL. The high degree of consistency between the laser and acoustic measurements show that either sensor is viable for net relative navigation. A fundamental advantage to the DVL is that it can measure velocity, but the laser triangulation sensor cannot.

The next steps include testing the laser sensor in closed-loop in an autonomous net following and mapping application.

4. Conclusions

We have shown experimentally that a laser triangulation can be used to navigate relative to an aquaculture net cage. The signal quality is nearly as good as a DVL, at less than 1/25th of the price.

Author Contributions: Software, H.B.A., S.J.O. and J.T.T.; data curation, W.C., S.J.O. and H.B.A.; formal analysis, J.T.T., M.B. and T.K. writing—original draft preparation, M.B., T.K.; writing—review and editing, M.B., T.K., W.C., H.B.A., S.J.O. and E.I.G. All authors have read and agreed to the published version of the manuscript.

Funding: The work is part of the project "SFI Exposed" (237790) with funding from the Research Council of Norway.

Institutional Review Board Statement: Ethical review and approval were waived for this study since appropriate operational measures were taken to minimize exposure of fish to laser radiation: (1) operate from inside the cage, (2) point constantly at the net form short distance, (3) turn on the laser only when necessary for the experiment itself.

Informed Consent Statement: Not applicable.

Data Availability Statement: The data presented in this study are available on request from the corresponding author.

Acknowledgments: Thanks to Gregory Bouquet for making the parallel laser illumination, and to Asbjørn Berge for fruitful discussions. Thanks, too, to Henrik Grønbech for helping with the modification of the Aqueous user interface.

Conflicts of Interest: The authors declare no conflict of interest.

References

1. Bianchi, M.C.G.; Chopin, F.; Farmer, T.; Franz, N.; Fuentevilla, C.; Garibaldi, L.; Grainger, N.H.R.; Jara, F.; Karunasagar, I.; Laurenti, A.L.G. *FAO: The State of World Fisheries and Aquaculture*; Food and Agriculture Organization of the United Nations: Rome, Italy, 2014.
2. Jensen, Ø.; Dempster, T.; Thorstad, E.; Uglem, I.; Fredheim, A. Escapes of fishes from Norwegian sea-cage aquaculture: Causes, consequences and prevention. *Aquac. Environ. Interact.* **2010**, *1*, 71–83. [CrossRef]
3. Føre, M.; Frank, K.; Norton, T.; Svendsen, E.; Alfredsen, J.A.; Dempster, T.; Eguiraun, H.; Watson, W.; Stahl, A.; Sunde, L.M.; et al. Precision fish farming: A new framework to improve production in aquaculture. *Biosyst. Eng.* **2018**, *173*, 176–193. [CrossRef]
4. Schjølberg, I.; Gjersvik, T.B.; Transeth, A.A.; Utne, I.B. Next Generation Subsea Inspection, Maintenance and Repair Operations. *IFAC-PapersOnLine* **2016**, *49*, 434–439. [CrossRef]
5. McLeod, D.; Jacobson, J.; Hardy, M.; Embry, C. Autonomous inspection using an underwater 3D LiDAR. In Proceedings of the 2013 OCEANS—San Diego, San Diego, CA, USA, 23–27 September 2013; pp. 1–8.
6. Bannister, J.; Sievers, M.; Bush, F.; Bloecher, N. Biofouling in marine aquaculture: A review of recent research and developments. *Biofouling* **2019**, *35*, 631–648. [CrossRef]
7. Matos, A.; Cruz, N.; Martins, A.; Pereira, F.L. Development and implementation of a low-cost LBL navigation system for an AUV. In Proceedings of the Oceans '99. MTS/IEEE. Riding the Crest into the 21st Century. Conference and Exhibition. Conference Proceedings (IEEE Cat. No.99CH37008), Seattle, WA, USA, 13–16 September 1999; Volume 2, pp. 774–779.
8. Alcocer, A.; Oliveira, P.; Pascoal, A. Study and implementation of an EKF GIB-based underwater positioning system. *Control Eng. Pract.* **2007**, *15*, 689–701. [CrossRef]
9. Rundtop, P.; Frank, K. Experimental evaluation of hydroacoustic instruments for ROV navigation along aquaculture net pens. *Aquac. Eng.* **2016**, *74*, 143–156. [CrossRef]
10. Rigby, P.; Pizarro, O.; Williams, S.B. Towards Geo-Referenced AUV Navigation Through Fusion of USBL and DVL Measurements. In Proceedings of the OCEANS 2006, Boston, MA, USA, 18–21 September 2006; pp. 1–6.
11. Cesar, D.B.D.S.; Gaudig, C.; Fritsche, M.; Dos Reis, M.A.; Kirchner, F. An evaluation of artificial fiducial markers in underwater environments. In Proceedings of the OCEANS 2015—Genova, Genoa, Italy, 18–21 May 2015; pp. 1–6.
12. Massot-Campos, M.; Oliver-Codina, G. Optical Sensors and Methods for Underwater 3D Reconstruction. *Sensors* **2015**, *15*, 31525–31557. [CrossRef] [PubMed]
13. Leone, A.; Diraco, G.; Distante, C. Stereoscopic System for 3-D Seabed Mosaic Reconstruction. In Proceedings of the 2007 IEEE International Conference on Image Processing, San Antonio, TX, USA, 16 September–19 October 2007; Volume 2, pp. 541–544.
14. Telem, G.; Filin, S. Photogrammetric modeling of underwater environments. *ISPRS J. Photogramm. Remote Sens.* **2010**, *65*, 433–444. [CrossRef]
15. Risholm, P.; Kirkhus, T.; Thielemann, J.T.; Thorstensen, J. Adaptive Structured Light with Scatter Correction for High-Precision Underwater 3D Measurements. *Sensors* **2019**, *19*, 1043. [CrossRef] [PubMed]
16. Risholm, P.; Thorstensen, J.; Thielemann, J.T.; Kaspersen, K.; Tschudi, J.; Yates, C.; Softley, C.; Abrosimov, I.; Alexander, J.; Haugholt, K.H. Real-time super-resolved 3D in turbid water using a fast range-gated CMOS camera. *Appl. Opt.* **2018**, *57*, 3927–3937. [CrossRef]
17. Jaffe, J. Computer modeling and the design of optimal underwater imaging systems. *IEEE J. Ocean. Eng.* **1990**, *15*, 101–111. [CrossRef]
18. Prats, M.; Fernandez, J.J.; Sanz, P.J. An approach for semi-autonomous recovery of unknown objects in underwater environments. In Proceedings of the 2012 13th International Conference on Optimization of Electrical and Electronic Equipment (OPTIM), Brasov, Romania, 24–26 May 2012; pp. 1452–1457.
19. Hildebrandt, M.; Kerdels, J.; Albiez, J.; Kirchner, F. A practical underwater 3D-Laserscanner. In Proceedings of the OCEANS 2008, Quebec City, QC, Canada, 15–18 September 2008; pp. 1–5.
20. Narasimhan, S.G.; Nayar, S.K.; Sun, B.; Koppal, S.J. Structured Light in Scattering Media. In Proceedings of the Tenth IEEE International Conference on Computer Vision (ICCV'05), Beijing, China, 17–21 October 2005; Volume 1, pp. 420–427.
21. Albiez, J.; Duda, A.; Fritsche, M.; Rehrmann, F.; Kirchner, F. CSurvey—An autonomous optical inspection head for AUVs. *Robot. Auton. Syst.* **2015**, *67*, 72–79. [CrossRef]
22. Moore, K.D.; Jaffe, J.S.; Ochoa, B.L. Development of a New Underwater Bathymetric Laser Imaging System: L-Bath. *J. Atmospheric Ocean. Technol.* **2000**, *17*, 1106–1117. [CrossRef]
23. Smith, R.C.; Cheeseman, P. On the Representation and Estimation of Spatial Uncertainty. *Int. J. Robot. Res.* **1986**, *5*, 56–68. [CrossRef]
24. Hidalgo, F.; Bräunl, T. Review of underwater SLAM techniques. In Proceedings of the 2015 6th International Conference on Automation, Robotics and Applications (ICARA), Queenstown, New Zealand, 17–19 February 2015; pp. 306–311.
25. Eustice, R.M.; Camilli, R.; Singh, H. Towards Bathymetry-Optimized Doppler Re-navigation for AUVs. In Proceedings of the Proceedings of OCEANS 2005 MTS/IEEE, Washington, DC, USA, 17–23 September 2005; pp. 1430–1436.
26. Roman, N.C. *Self Consistent Bathymetric Mapping from Robotic Vehicles in the Deep Ocean*; Massachusetts Institute of Technology: Cambridge, MA, USA, 2005.

27. Kinsey, J.C.; Eustice, R.M.; Whitcomb, L.L. A survey of underwater vehicle navigation: Recent advances and new challenges. In Proceedings of the IFAC Conference of Manoeuvering and Control of Marine Craft, Girona, Spain, 17–19 September 1997; Volume 88, pp. 1–12.
28. Amundsen, H.B. Robust Nonlinear ROV Motion Control for Autonomous Inspections of Aquaculture Net Pens. Master's Thesis, Department of Engineering Cybernetics, Norwegian University of Science and Technology (NTNU), Trondheim, Norway, 2020.
29. Caharija, W.; Pettersen, K.Y.; Bibuli, M.; Calado, P.; Zereik, E.; Braga, J.; Gravdahl, J.T.; Sorensen, A.J.; Milovanovic, M.; Bruzzone, G. Integral Line-of-Sight Guidance and Control of Underactuated Marine Vehicles: Theory, Simulations, and Experiments. *IEEE Trans. Control Syst. Technol.* **2016**, *24*, 1623–1642. [CrossRef]
30. Candeloro, M.; Sørensen, A.J.; Longhi, S.; Dukan, F. Observers for dynamic positioning of ROVs with experimental results. *IFAC Proc. Vol.* **2012**, *45*, 85–90. [CrossRef]
31. Shortis, M. Camera Calibration Techniques for Accurate Measurement Underwater. In *3D Recording and Interpretation for Maritime Archaeology*; McCarthy, J.K., Benjamin, J., Winton, T., van Duivenvoorde, W., Eds.; Springer International Publishing: Cham, Switzerland, 2019; pp. 11–27.
32. Zhang, Z. A flexible new technique for camera calibration. *IEEE Trans. Pattern Anal. Mach. Intell.* **2000**, *22*, 1330–1334. [CrossRef]
33. Torr, P.H.S.; Zisserman, A. MLESAC: A New Robust Estimator with Application to Estimating Image Geometry. *Comput. Vis. Image Underst.* **2000**, *78*, 138–156. [CrossRef]

Article

Towards Multi-Robot Visual Graph-SLAM for Autonomous Marine Vehicles

Francisco Bonin-Font † and Antoni Burguera *,†

Systems, Robotics and Vision Group, Department of Mathematic and Informatics, University of the Balearic Islands, 07122 Palma, Spain; francisco.bonin@uib.es

* Correspondence: antoni.burguera@uib.es

† These authors contributed equally to this work.

Received: 16 May 2020; Accepted: 10 June 2020; Published: 14 June 2020

Abstract: State of the art approaches to Multi-robot localization and mapping still present multiple issues to be improved, offering a wide range of possibilities for researchers and technology. This paper presents a new algorithm for visual Multi-robot simultaneous localization and mapping, used to join, in a common reference system, several trajectories of different robots that participate simultaneously in a common mission. One of the main problems in centralized configurations, where the leader can receive multiple data from the rest of robots, is the limited communications bandwidth that delays the data transmission and can be overloaded quickly, restricting the reactive actions. This paper presents a new approach to Multi-robot visual graph *Simultaneous Localization and Mapping* (SLAM) that aims to perform a joined topological map, which evolves in different directions according to the different trajectories of the different robots. The main contributions of this new strategy are centered on: (a) reducing to hashes of small dimensions the visual data to be exchanged among all agents, diminishing, in consequence, the data delivery time, (b) running two different phases of SLAM, intra- and inter-session, with their respective loop-closing tasks, with a trajectory joining action in between, with high flexibility in their combination, (c) simplifying the complete SLAM process, in concept and implementation, and addressing it to correct the trajectory of several robots, initially and continuously estimated by means of a visual odometer, and (d) executing the process online, in order to assure a successful accomplishment of the mission, with the planned trajectories and at the planned points. Primary results included in this paper show a promising performance of the algorithm in visual datasets obtained in different points on the coast of the Balearic Islands, either by divers or by an *Autonomous Underwater Vehicle* (AUV) equipped with cameras.

Keywords: multi robot; Simultaneous Localization and Mapping; visual loop closure; image global signatures

1. Introduction and Related Work

Simultaneous Localization and Mapping (SLAM) [1] is an essential task for *Autonomous Underwater Vehicles* (AUV) to achieve successfully and precisely their programmed missions. SLAM consists of building a map of the environment and, at the same time, estimating its own pose within this map. SLAM is presently a de facto localization standard for any kind of autonomous vehicle. Laser range finders or sonar were the sensor modality of choice at first [2–4]. However, research turned to computer vision as soon as price and capabilities of cameras made it possible [5], since cameras provide higher temporal and spatial data resolutions and richer representations of the world.

However, large-scale or long-term operations with a single robot equipped with cameras generate huge amounts of visual data that can collapse the vehicle computer, if they are not treated properly. A common strategy to overcome this problem is to explore the areas of interests in different, separated missions, so-called *sessions*, run with a single robot in different time periods (a Multi-session configuration [6]) or with several robots running simultaneously (Multi-robot configurations [7]). Therefore, any low capability of a robot to operate robustly during long periods of time can be alleviated by running different transits with different agents, at the same time through common areas, and joining all individual estimated trajectories in a single coordinate frame. Multi-robot systems also increase robustness in case of failure of any of the robots; however, they need complex coordination and multiple localization systems. Typical applications using teams of robots include aerial surveillance [8], underwater exploration [9], maintenance of industrial infrastructures or intervention in archaeological sites [10], among others.

The first approaches to Multi-robot SLAM were based on particle filters [11], and introduced the concept of *encounters* as the relative pose between two robots that can mutually recognize each other and determine their relative poses. These *encounters* are introduced as additional pose constrains in the particle filter. Some Multi-robot approaches are based on the *Anchor-nodes* [12,13] proposal, which defined two concepts unconsidered for multiple trajectories until that moment: (a) the *Anchor*, defined as the offset of a complete trajectory with respect to a global system of coordinates, and (b), an *encounter*, re-defined as a transformation between two different poses of two different robots that observe the same part of the environment, but without being necessarly that both robots recognize themselves. In visual-based systems this can be achieved, for instance, detecting overlapping scenes. In Multi-robot systems, *encounters* represent additional constraints between different graphs corresponding to different sessions.

Schuster et al. conceived a very precise SLAM approach to localize a team of planetary rovers equipped with an *Inertial Measurement Unit* (IMU) and a stereo camera [14]. IMU data, visual odometry and wheel odometry are integrated in a local localization *Extended Kalman Filter* (EKF) and the 3D point clouds of all robots computed from their respective stereo views are, firstly stored in each agent, and then matched to be joined in global 3D maps. The use of stereo vision and advanced techniques for 3D feature matching and alignment complicate considerably the whole system and generate huge amounts of data to work with and to be exchanged. This solution turns out to be very difficult for underwater missions, given the limited options for fast communication in this media.

Saeedi et al. offered an extensive survey of Multi-robot systems and strategies, pointing also towards the upcoming trends and challenges [15], such as extending the systems to dynamic and/or large-scale environments or increasing the number of agents in the working teams.

Another issue to consider in SLAM is the detection of loop closings and their use to correct the robot trajectory estimated by means of dead-reckoning sensors, such as, inertial units, acoustic beacons, laser-based or visual odometers. Loop closing is the problem of recalling revisited scenes, and approaches to visual loop closure detection try to recognize the same scene in different images, taken at different and relatively distant time instants, regardless evident differences on scale or view point [16]. In single session SLAM, since the robot pose is continuously estimated, the search for images candidate to close a loop with a *query* (from now on called intra-session loop closings) is constrained to a region around the robot pose associated with that query [17]. In contrast, multi-robot loop detection, i.e., the detection of loop closings among different sessions of different robots (from now on called inter-session loop closings), cannot rely on the AUV poses to constrain the search since, at first, the relative poses between sessions is unknown. Consequently, it seems that every *query* of one session would need to be compared with all the images obtained until that moment in the other sessions, increasing considerably the time dedicated for this task, and the amount of visual data to be exchanged.

Exchanging image hashes instead of entire images or sets of image salient points is a way to reduce data transfer requirements in Multi-robot configurations. Hash functions are usually used to authenticate messages sent between a source and a receiver, so that the later can verify the authenticity of the source. Conventional hashing algorithms are extremely sensitive to the hashed messages. A change in 1 bit of the input message causes dramatic changes on the output. Applications of hashes, understood as exposed before, include image retrieval in large databases, authentication and watermarking, among many others [18]. However in applications of scene recognition, localization or visual loop closing detection, it is accepted that similar or overlapping images are expected to produce similar or close hashes while distinct images produce clearly distinctive hashes, being this concept known are perceptual image hashing [19–23]. In particular, McDonald et al. [6] proposed to detect loop closings using a solution based on *Bag of Words* BoW [24] combined with iSAM [12] for batch map optimization, and Negre et al. [22] showed how their new global image descriptor HALOC outperformed other techniques, such as BoW and VLAD [20], in the task of loop closing detection with image hashes. From now on, this text uses equally hash or global image descriptor to refer the same concept.

All these aforementioned references apply hashes to detect loops in SLAM applications for single robots. However, now, our interest is focused on the Multi-robot systems, and the application of global image signatures to find loops between images captured by different robots that operate in a same mission on a common area of interest. A few authors have already explored this idea. For instance, *Decentralized Visual Simultaneous Localization and Mapping* (DSLAM) [25] is a powerful tool for pose-graph Multi-robot decentralized applications in environments where absolute positioning is not available. DSLAM reduces every image to its NetVLAD (a Neural Network Architecture for Place Recognition) global descriptor [26]. To find loop closings, DSLAM seeks, for every *query* of one robot, the image of another robot whose NetVLAD descriptor presents the shortest distance to the descriptor of the *query* and this distance is below a certain threshold. This process is done for every frame of every robot that is inside a predefined cluster. DSLAM uses ORB-SLAM [27] for continuous localization, which includes a global image charectrization based on BoW for initial odometric estimates, and ORB [28] feature matching and RANSAC to calculate the 3D transform between confirmed visual loop closings.

The idea of *Cloud Computing* is applied in some cases to alleviate the computational charge needed for a set of robots to localize themselves and map the environment running a software architecture based on a multi-layer cloud platform [29]. In this later reference, robots use ORB-SLAM for self-localization and the multi agent SLAM is tested with the KITTI [30] public dataset and using a quadrotor drone in an outdoor environment. A few solutions integrate inertial with image data to perform Multi-robot graph SLAM. In [31], ORB visual features are tracked along consecutive frames and integrated together with the motion given by an IMU in a graph optimization context. BoW is also used to detect candidates to close inter-session loops. The BoW-based global image descriptor of a query image is compared to the global descriptor of all other images of the other agents, selecting a set of candidates to close inter-session loops with the query. Afterwards, a brute-force feature matching with RANSAC is applied to confirm the candidates or to reject them. Experiments in [31] are performed with aerial robots in industrial environments.

Previous references have been tested only in terrestrial indoor and outdoor environments. The literature is extremely scarce in Multi-robot SLAM addressed, implemented and tested in underwater scenarios with AUVs [9,32]. Underwater computer vision is affected by several challenging problems, such as flickering, reduced range, lack of illumination, haze, light absorption, refraction, and reflection. These limitations increase the need for more robust visual SLAM approaches which start with accurate camera calibrations. Accuracy in the processes of camera calibration is critical to reduce drift in the visual odometry and increase precision in the pose transform obtained from images that close loops [33,34]. Furthermore, none of the papers cited previously consider the potential impact of limited communications among robots, because, either they are applied in ground or aerial robots or they simply assume full, high-bandwidth connectivity. This supposal is clearly unrealistic in underwater environments, where blue-light laser communications need to be highly directive and acoustic USBL modems work, on average, at 13 Kbps for long range devices and up to 65 Kbps in mid-range devices. Additionally, the later speeds are not suitable to transmit medium-high resolution visual data between two robots without a previous compression. For instance, Pfingsthorn et al. [35] proposed a visual pose-graph SLAM approach in which compressed JPEG-format images are send via acoustic links only between robots that can mutually recognize their positions and are viewing overlapping areas. Paull et al. [36] refuse the use of images for SLAM and trust all the localization process to an acoustic modem for data transmission and instruments that give relatively small amounts of data, if compared with images: compass, *Doppler Velocity Log* (DVL) and a Side Scan Sonar. Besides, they also apply a new strategy to marginalize unnecessary local information to reduce the dimensions of the transferable packages.

In the context of the ongoing national project TWINBOT (*TWIN roBOTs for Cooperative Underwater Intervention Missions*) [37], diverse missions of exploration and cooperative intervention have to be run using one or several AUVs in underwater areas with multiple appearances and different benthic habitats. In this project, accurate, fast and reliable robot localization, loop closing and navigation algorithms are crucial for the success of their missions. This paper presents a new approach to Multi-robot visual graph-SLAM, especially designed for 2′5D configurations, where vehicles move at a constant altitude with a camera pointing downwards, with the lens axis constantly perpendicular to the ground or to the vehicle longitudinal axis. This condition simplifies the visual system to 3 *Degrees of Freedom* (DoF): two for an in-plane translation (x, y) and another for rotation in yaw (θ). This simplification fits with aerial and underwater vehicle configurations, if the navigation altitude is large enough compared with the heigh of the terrain relief [38,39]. However, now tests have been made only with underwater datasets since the research developed by our team, in general [40], and the TWINBOT project in particular, is applied entirely and solely underwater, and this approach emerged as a solution to be applied on the robots that participate in our project missions.

The approach presented now includes several contributions that represent clear advantages with respect to the existing solutions, namely:

(a) As in [25], images are reduced to global descriptors decreasing drastically the amount of visual data to be exchanged among robots; however, the global descriptor used now is HALOC [22] instead of NetVLAD. The construction of HALOC is simpler and faster than NetVLAD, consisting in projecting all image features on a base of orthogonal vectors, without any need for tedious and long training tasks. This is a clear advantage over the previous work, since HALOC already showed to outdo VLAD [20] and BoW, in speed and performance for loop closing detection, in both terrestrial public benchmaks and underwater environments. HALOC also showed a performance better than ORB-SLAM, only underwater. Additionally, extensive experiments with HALOC performed in marine areas partially colonized with seagrass [41,42] also revealed an excellent efficiency, capacity and utility for loop closing detection in this type of environments.

(b) A second important contribution is the simplification of the whole system with respect previous approaches. Ours does not require neither the computation of relative poses among robots, nor a specific strategy to limit their communication and interaction. At every SLAM iteration, the quantity of bytes to be exchanged between robots is so small that this will not necessarily limit the communication between all agents that participate in the mission, if needed.

(c) The global procedure includes local and global SLAM tasks, with a map joining process in between. The advantage of this point lies on the flexibility to choose the moment at which the map joining is performed, giving priority to local routes as accurate as possible, or delaying the major corrections once all maps have been joined.

(d) The present approach goes one step beyond its predecessors, since the joined graph incorporates and reflects, online, the successive poses of all robots that move simultaneously.

(e) The localization and motion problem is simplified to 2D. Furthermore, it avoids complex multi-layer software architectures or Cloud computing strategies present in previous solutions.

(f) One of the principal objectives has been the reduction of the computational requirements of the algorithm, since, in general, they are limited in lightweight underwater vehicles. Running the algorithm online onboard the vehicles is a must, since it is especially addressed to multi-robot configurations, and these configurations imply controlling, mapping and guiding several robots moving simultaneously, where usually, one centralizes the processing of the localization data of the whole group.

Furthermore, although they are not directly novel contributions, it is worth mentioning two additional advantages in the implementation: (i) similarly to [32] or [35], once maps of different robots are joined, standard graph-based topology representations are used, where images form nodes and transforms between two images (being from consecutive frames or between two images that close a loop) form edges or links, and (ii) the graph is optimized by means of standard *g2o* [43]; this standardization facilitates the exchange of the different modules on a variety of software platforms and their reuse among different implementations.

Although this is out of the scope of this paper, this vision-based algorithm can complement the navigation facilities of underwater vehicles equipped with multiple types of sensors. In fact, this algorithm can integrate additional sensorial data in the first estimation of the vehicle motion, combining visual odometry with other means of laser or sonar-based dead reckoning [44,45].

The source code of the whole approach has been made publicly available for the scientific community in several GitHub repositories, together with a simple underwater dataset to test the whole procedure. Links to sources are provided in Section 3.

Section 2 contextualizes and details all algorithms proposed to: (a) estimate the visual odometry, (b) detect intra- and inter-session loop closings, (c) perform the local trajectory-based SLAM and (d) join maps and optimize the global graph. Section 3 presents some qualitative and quantitative preliminary results. Finally, Section 4 concludes the paper and gives some indications of ongoing and upcoming tasks to continue and improve this work.

2. Materials and Methods

2.1. Overview

The proposed localization module is based only in vision, with no intervention of either dead-reckoning navigation instruments, such as IMU or DVL, or global positioning systems, such as GPS for surface vehicles or Ultra-short Baselines (USBL) for underwater vehicles.

The structure of the proposed system is as follows:

(1) Let us simplify the problem assuming that there are, for instance, two vehicles moving simultaneously over the same area of interests in such a way that there is no possibility of collision, and that part of the area will be explored by both robots.

(2) The approach starts by estimating the trajectory of each robot motion, separately, applying the trajectory-based visual-SLAM strategy included in the multi-session scheme of Burguera and Bonin-Font [46]. Let us refer to this step as the *intra-session SLAM*. The indicated trajectory-based scheme implies that the trajectory of each robot is estimated by means of compounding [47] successive displacements calculated from one point to the next. These successive displacements form the state of an Extended Kalman filter (EKF) which is updated using the transforms given by the confirmed loop closings. In our particular case, this displacement corresponds to the visual odometry calculated between consecutive images, and the images candidate to close a loop with the current image are found comparing the corresponding image hashes and confirmed by a RANSAC-based algorithm applied on a brute-force visual feature-matching process. The main difference with respect to [46] is that while in a multi-session localization procedure the trajectory of the currently running robot is joined to another trajectory already completed and available in its totality, now, in a Multi-robot scheme, both robots are moving at the same time to complete a mission in which both participate simultaneously. When joined, both trajectories are incomplete, and continue running until all robot missions are finished.

(3) Simultaneously to both *intra-session SLAM* tasks, the system also searches for inter-session loop closings using HALOC. The global signature of each new image captured by each robot is compared with the global signatures of all images of the other robot. Once a certain number of inter-session

loop closings are confirmed, both routes are joined in a single graph. Let us refer to this step as the *Map Joining*.

(4) Once joined, the global map (graph) must be completed with the successive poses of both robots until the end of both sessions. Furthermore, the trajectory-based localization approach applied to both agents separately is not longer valid. Each new displacement of both robots is included in the form of new nodes on the graph. Each new node of the graph will follow the direction of motion of each vehicle, which means that the graph will grow in two different directions, according to the two different trajectories, but forming a single entity. Let us denote this step as the *Multi-robot SLAM*.

(5) The global graph completion and optimization is done using a pose-based scheme, i.e., all new nodes corresponding to each robot will contain their successive global poses with respect to the origin of a unique world coordinate system, while the links between nodes will contain the displacement between them. After the map joining, only the inter-session loop closings are used to optimize the global map. If one loop closing between different sessions is confirmed, its resulting transform is used as an additional constraint between two nodes to optimize the whole graph. In our case, the optimization solver used is the *g2o* implementation [43] of the *Levenberg Marquardt* algorithm [48,49].

2.2. Intra-Session SLAM and Map Joining

2.2.1. Visual Odometry

Figure 1 shows the global idea behind this first step of the approach. The visual odometry gives the estimated 2D motion between consecutive images by means of a SIFT feature detection and matching procedure.

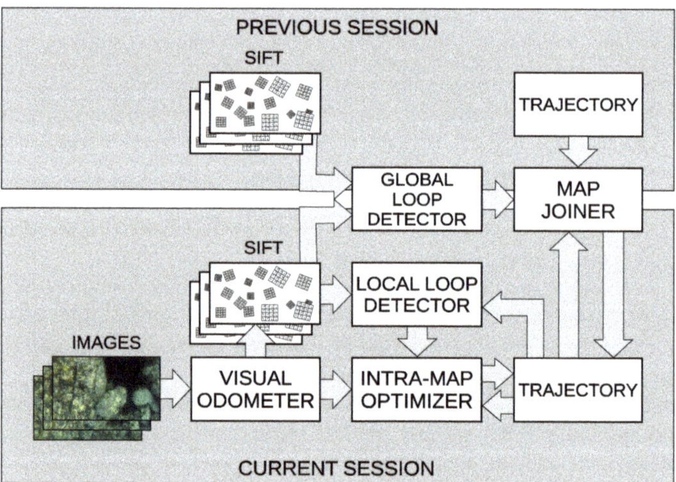

Figure 1. Sigle Session SLAM overview.

Algorithm 1 shows the RANSAC-based method used to register two images, i.e., get the transform (if it exists) between them in translation and rotation. *apply_altitude*() is a function that converts image feature coordinates from pixels to meters by considering the altitude at which the AUV navigates, as well as the camera parameters.

Algorithm 1: RANSAC approach to estimate the motion $\hat{X}_B^A$ from image I_A to image I_B.

Input:
f_A, f_B: SIFT features in images I_A and I_B
a_A, a_B: Altitudes corresponding to I_A and I_B
C: Set of correspondences
K: Number of iterations to perform
N_{corr}: Number of correspondences to be randomly selected
N_{min}: Minimum number of correspondences to consider a roto-translation as candidate
ε_{corr}: Maximum allowable error per correspondence
Output:
$fail$: Boolean stating if failed to find $\hat{X}_B^A$
$\hat{X}_B^A$: The estimated roto-translation

begin
  $f'_A \leftarrow apply\_altitude(f_A, a_A)$;
  $f'_B \leftarrow apply\_altitude(f_t, a_B)$;
  $\varepsilon_B^A \leftarrow \infty$; $fail \leftarrow true$;
  for $i \leftarrow 0$ **to** $K-1$ **do**
    $R \leftarrow$ random selection of N_{corr} items from C;
    $X \leftarrow \arg\min_T \sum_{(i,j)\in R} ||T \oplus f'_{A,i} - f'_{B,j}||$;
    $\varepsilon \leftarrow \sum_{(i,j)\in R} ||T \oplus f'_{A,i} - f'_{B,j}||$;
    foreach $(i,j) \in (C - R)$ **do**
      if $\|X \oplus f'_{A,i} - f'_{B,j}\| < \varepsilon_{corr}$ **then**
        $R \leftarrow R \cup \{(i,j)\}$;
      end
    end
    if $|R| > N_{min}$ **then**
      $X \leftarrow \arg\min_T \sum_{(i,j)\in R} ||T \oplus f'_{A,i} - f'_{B,j}||$;
      $\varepsilon \leftarrow \sum_{(i,j)\in R} ||T \oplus f'_{A,i} - f'_{B,j}||$;
      if $\varepsilon < \varepsilon_B^A$ **then**
        $\varepsilon_B^A \leftarrow \varepsilon$; $\hat{X}_B^A \leftarrow X$; $fail \leftarrow false$;
      end
    end
  end
end

The idea behind this algorithm is that correct correspondences lead to the same roto-translation while wrong feature matchings lead to different and wrong roto-translations. The algorithm selects a random subset R of correspondences from the total number of correspondences C between two images, and then computes the roto-translation X and the subsequent error ϵ using only this subset. Afterwards, if the error introduced by the non-selected matchings of C is below a threshold ϵ_{corr}, then, these matchings are included in R. If at any moment, the number of elements in R surpasses a threshold N_{min}, the roto-translation and the error are computed again using this expanded R. If the error is below the smallest error obtained until this moment, the roto-translation is kept as a good model. This process is iterated a certain number of times. If partial roto-translations are inconsistent and R never reaches the minimum number of items required, the algorithm will not return any transform, but a boolean called $fail$ set to $true$. The obtained transform can be assumed to be the odometric displacement between consecutive images and the trajectory of the robot between steps i and j (X_j^i) (assuming step j being subsequent to i) can be estimated using the compounding $\oplus$ operator of the successive odometric displacements ($X_{i+1}^i, X_{i+2}^{i+1}, \ldots, X_j^{j-1}$), as described in [50]:

$$X_j^i = X_{i+1}^i \oplus X_{i+2}^{i+1} \oplus \cdots \oplus X_j^{j-1} \quad j > i \ . \tag{1}$$

2.2.2. Local Loop Detection and Trajectory Optimization

Local loops are those found within a single SLAM session. The Loop Candidates set (LC_t) is the set of images I_i that may close a loop with the last gathered image I_t obtained in each running trajectory. This set is built by searching in a region within a predefined radius δ [38] around the current robot pose as estimated by the odometry:

$$LC_t = \{i : ||X_t^i||_2 \leq \delta, i < t-1\} \quad (2)$$

where X_t^i is computed by Equation (1).

Every image contained in the set of loop closing candidates ($I_i \in LC_t$) is registered with I_t using Algorithm 1, in order to build the set of local loops LL_t, being $LL_t = \{Z_t^i : i \in LC_t \cap \neg fail(i,t)\}$, where $\neg fail(i,t)$ indicates that Algorithm 1 did not fail to get a roto-translation Z_t^i between I_i and I_t.

The trajectory estimation obtained by means of compounding the successive odometric displacements between two points A and B will most likely not coincide with the direct transform between images obtained in A and B provided by the image registration process of Algorithm 1, if A and B close a loop:

$$X_{A+1}^{A} \oplus X_{A+2}^{A+1} \oplus X_{A+3}^{A+2} \oplus \cdots \oplus X_{B-1}^{B-2} \oplus X_{B}^{B-1} \neq Z_{B}^{A} \quad (3)$$

due to the drift introduced by the visual odometry and the error inherent to the transform directly obtained from the image registration procedure. Figure 2 illustrates these concepts.

Afterwards, a process of global optimization is run to get a trajectory that best combines the pose constraints imposed by the set of local loops (LL_t) and the odometry. As mentioned before, the trajectory of the robot is the state vector of a *Iterative Extended Kalman Filter* (IEKF). Each new odometric displacement (X_t^{t-1}) computed between the last image and the previous one is used to augment the state vector at time t (X_t^-): $X_t^- = \begin{pmatrix} (X_{t-1})^T & (X_t^{t-1})^T \end{pmatrix}^T$. In the prediction stage of the IEKF the state vector does not change at all.

If LL_t is not empty, the trajectory is optimized performing the Update stage of the IEKF using the set of loop closings as measurements. The observation function h_t^i associated with each measurement Z_t^i (the transform of each real loop closing) can be defined as $h_t^i(X_t^-) = X_{i+1}^i \oplus X_{i+2}^{i+1} \oplus \cdots \oplus X_{t-1}^{t-2} \oplus X_t^{t-1}$, being the innovation of the IEKF for each measurement: $Z_t^i - h_t^i(X_t^-)$. With all this elements, one can iterate the classical equations of an EKF to get the optimized trajectory, until $||X_{t,j}^- - X_{t,j-1}^-|| < \gamma$, where j represents the last iteration and γ is a predefined threshold. The classical format of the IEKF involves iterating until the changes between consecutively estimated states are below a certain threshold. However, in the experiments we verified that after a certain and almost constant number of iterations, the filter already converged with a difference between consecutive results below the threshold ($||X_{t,j}^- - X_{t,j-1}^-|| < \gamma$). Because of that, it was decided to repeat all the experiments with a defined number of iterations, in order to limit the number of executions to be done and save computational resources. In any case, both options can be used in other circumstances, depending on the needs and environmental conditions of each different system and field case.

2.2.3. Inter-Session Loop Closings

The main problem for joining two trajectories of two robots operating simultaneously is the lack of geometric relation between their corresponding sessions. Every robot geo-localizes itself with respect to the origin of its own trajectory, but it has no knowledge about the origin of the other trajectories. By means of finding inter-session loop closings, i.e., images that show, partially or totally, the same area, but taken by two different robots in two different sessions running simultaneously, the maps of the two robots can be joined in a single one [46]. Due to this lack of geometrical relation between the two trajectories, the search of the loop closing candidates of one robot to close a loop with the last image captured by the other robot cannot be restricted to a certain area. In this case, one should compare the last image gathered by one of the robots with all images taken by the other robot, from the start of the mission until the current moment. Applying a brute force feature matching algorithm between all these involved images is unfeasible for online applications, due to the great amount of computing resources and time needed. One way to alleviate this problem is reducing all images to global descriptors. As in [46], all images of both sessions are reduced to their HALOC global descriptor. The size of HALOC is fixed in 384 floats since the size of the used projective orthogonal vectorial space is 3 [22]. This length is independent of the number of visual features found in each image. The global descriptor of every new image of one of the sessions (called the query image) is compared with the global descriptor of all and each of the images taken during the other session. According to [22], those 5 images that give the lowest L1-norm of the difference between their hash and the query hash, and this norm is lower than a certain threshold δ' are taken as the inter-session loop closing candidates (GC_t):

$$GC_t = \{i : ||H_i - H_t||_1 \leq \delta', \forall I_i \in V_p\} \quad (4)$$

being V_p the set of images taken by one of the robots from the start of its session until the current moment, H_i the hash of each of these images and H_t the hash of the query image. The value of δ' will be selected experimentally.

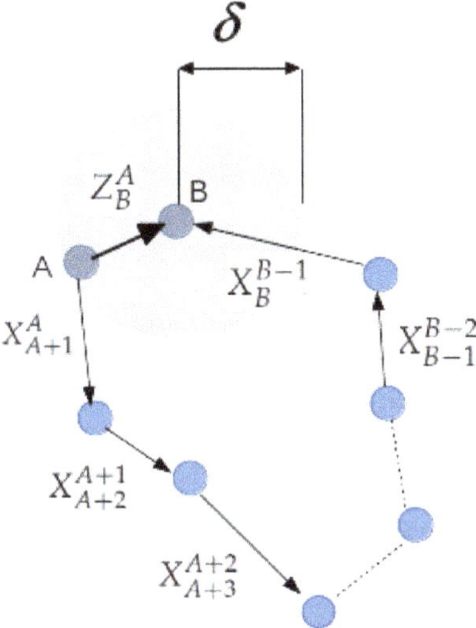

Figure 2. In theory, the transform between A and B, if both close a loop should be very close to the transform obtained compounding the odometric displacements X_{i+1}^i.

Once the set of candidates is established, the true positives are confirmed by means of the RANSAC-based Algorithm 1, forming the definitive set of images (GL_t) of the first session that, in principle, close a loop with the last image of the second session, as: $GL_t = \{Z_t^i : i \in GC_t \cap \neg fail(i,t)\}$, being Z_t^i the transformation found by Algorithm 1. Inter-session loops are accumulated at every iteration of both single SLAM sessions. These transforms Z_t^i, are, in fact, a set of geometrical relations between the two different sessions. Assessing the performance of HALOC in loop closing detection, in terms of accuracy, recall and fall-out, is out of the scope of this paper, since it has already been presented in [22,42] with considerable good results underwater.

2.2.4. Map Joining

As mentioned in the previous section, the loop closings between different sessions can be used to infer the geometrical relation between the two trajectories of the two robots that perform both missions simultaneously. The objective now is to align, at a certain moment, both surveys in a single global graph, and, maintain this single graph from the moment of the joining to the end of both missions.

Let X_1 denote the trajectory of one of the sessions. Let the first and last images of this trajectory be denoted as I_{1s} and I_{1e}, respectively. Let X_2 denote the trajectory of the second session and let us denote its first and last images as I_{2s} and I_{2e}, respectively. Let us denote the number of accumulated inter-session loop closings at a certain moment t as K. Let us also denote this set of loop closings as Z_G, each one relating one image of the first session with another image of the second session.

$$Z_G = \left((Z_{20}^{10})^T \quad (Z_{21}^{11})^T \quad \dots \quad (Z_{2K-1}^{1K-1})^T \right)^T \tag{5}$$

where each Z_{2i}^{1i} represent a transform from image I_{1i} of the session 1 to image I_{2i} of session 2, or what is the same, the transforms of the loop closings. Each Z_{2i}^{1i} belongs to a certain GL_t.

Let us define X_{2s}^{1e} as the transform, or the relative motion, from I_{1e} to I_{2s}. For every loop closing, ideally, $Z_{2i}^{1i} = X_{1e}^{1i} \oplus X_{2s}^{1e} \oplus X_{2i}^{2s}$, where X_{1e}^{1i} is the displacement from I_{1i} to I_{1e}, and X_{2i}^{2s} is the displacement from I_{2s} to I_{2i}, being:

$$X_{1e}^{1i} = (x_{1e}^{1i}, y_{1e}^{1i}, \theta_{1e}^{1i})^T \tag{6}$$

$$X_{2s}^{1e} = (x_{2s}^{1e}, y_{2s}^{1e}, \theta_{2s}^{1e})^T \tag{7}$$

$$X_{2i}^{2s} = (x_{2i}^{2s}, y_{2i}^{2s}, \theta_{2i}^{2s})^T \tag{8}$$

The proposal consists of an IEKF that will give the value of X_{2s}^{1e} that better matches all the loop closures found until the moment t. The state vector of the IEKF is just the transform X_{2s}^{1e}, the observation function for each loop closing will be $g_G^i = X_{1e}^{1i} \oplus X_{2s}^{1e} \oplus X_{2i}^{2s}$, and Z_{2i}^{1i} is the corresponding measurement. With this, one can form the innovation, and apply the classical EKF equations iteratively, as explained in Section 2.2.2. X_{2s}^{1e} is the transformation that can be used to join the two sessions, in such a way that the state vectors of both trajectories, formed by displacements, are joined by this recently computed transformation as: $X_J = \left((X_1)^T (X_{2s}^{1e})^T (X_2)^T\right)^T$, where X_J represents the joined trajectory, and $X_1 // X_2$ the state vector of the first and second trajectories, respectively, from their starting points until the instant t. The idea is illustrated in Figure 3.

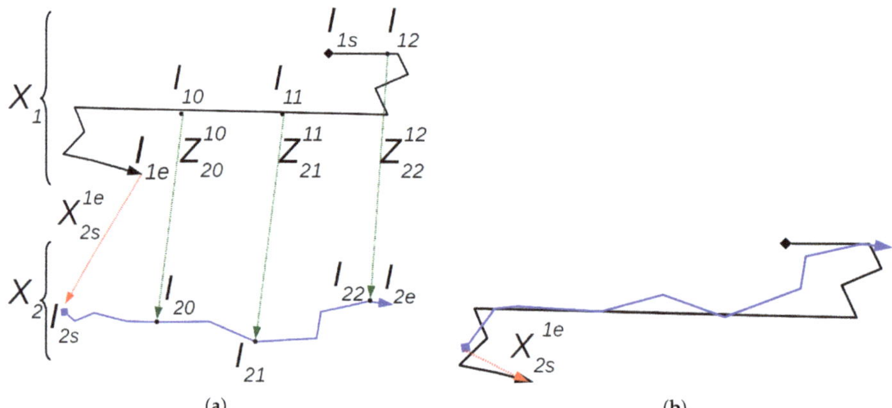

Figure 3. (**a**) Separated trajectories with intersession loop closings. (**b**) The joined trajectory.

Once both sessions have been joined, the trajectory-based schema is no longer valid, and the resulting map is transformed into a pose-based graph. All the robot displacements included in X_J are transformed into global poses that constitute each node of the global graph as:

$$X^i = \begin{cases} X_1^0 \oplus X_2^1 \oplus \cdots \oplus X_{i-1}^{i-2} \oplus X_i^{i-1} & i <= 1e \\ X_1^0 \oplus X_2^1 \oplus \cdots X_{1e}^{1e-1} \oplus X_{2s}^{1e} \oplus X_{2s+1}^{2s} \oplus \cdots \oplus X_i^{i-1} & i > 1e \end{cases} \tag{9}$$

where X^i is the pose associated with node i. The case for $i <= 1e$ refers to the global pose of a node corresponding to an element of the first trajectory, and the case for $i > 1e$ refers the global pose of a node of the second trajectory. All the displacements included in X_J become the links between successive nodes and each node of the graph is associated with its corresponding image.

2.3. Multi-Robot Graph SLAM

Let us assume that, (a) after both trajectories have been joined in a single graph, both robots are still running their own missions, and, (b) the successive poses of both robots must be included in the global graph as new nodes, each one following the corresponding trajectory.

The Multi-robot Graph-SLAM procedure detailed below follows the indications of [51], in terms of structure, node generation, inclusion of loop closings as additional pose constraints, and graph optimization, but in our case particularized for a Multi-robot configuration. The algorithm includes the next points:

(1) The local SLAM algorithm explained in Section 2.2 is continuously executed for both trajectories until them are joined when a certain number of inter-session loop closings have been accumulated. It is better to optimize the local trajectories every N frames, although there is only a couple of loop closings, in order to, when both sessions are joined the drift has already been reduced locally, as much as possible. Otherwise, trajectories could be joined before local optimizations have been applied, transferring local drifts to the global map.

Let us assume that the Multi-robot localization is centralized in the first robot, which will receive, from the second robot: (1) The set of visual features and the global descriptor of the last gathered image, (b) only if the map joining has to be done, the state vector and the last odometric displacement. The state vector is needed to be attached to the one of the first robot, if it is due. The last frame global descriptor is needed to find possible loop closures with frames of trajectory 1, the set of features is needed to confirm or reject the possible candidates, and the last odometric displacement of trajectory 2 will be used as a reference after the map joining.

(2) Once both trajectories have been joined in a single global graph, it is time to feed the map with the successive displacements of both robots. It is important to note that the last node of the graph corresponds to the last displacement of the second trajectory, since the first set of elements correspond to the trajectory of robot 1, then it comes the link between trajectories, and finally the elements of trajectory 2. Let us denote the identifier ID of the last node of the global graph as N_{n2}, where $n2$ represents the number of nodes in the graph, and is equal to the length of the joined state vector X_J ($|X_J|$). Accordingly, the ID of the graph node corresponding to the end of the trajectory 1 will be N_{n1}, where $n1 = |X_J| - 1 - |X_2|$.

Storing N_{n1} and N_{n2} is necessary, since they will be the points of the global graph from which the successive nodes corresponding to trajectories 1 and 2, respectively, will be placed according to the ongoing motion of both vehicles.

The set of iterated actions performed for the Multi-session SLAM are:

1. Let us denote the last (or next) computed odometric displacements of trajectories 1 and 2 as X_{n1} and X_{n2}, respectively. These displacements together with the last images of both session are stored in the system.
2. If trajectory 1 has not finished, add a new node (N_{n1+1}) to the graph, linked to N_{n1} with the transform X_{n1}. The global pose contained in this node will be: $X^{n1+1} = X^{n1} \oplus X_{n1}$. N_{n1+1} will be the last node of trajectory 1.
3. If the trajectory 2 has not finished, add a new node (N_{n2+1}) to the graph, linked to N_{n2} with the transform X_{n2}. The global pose of this node will be: $X^{n2+1} = X^{n2} \oplus X_{n2}$. N_{n2+1} will be the last

node of trajectory 2. The link between nodes X^{n1+1} and X^{n1}, and the link between nodes X^{n2+1} and X^{n2} will contain the values of X_{n1} and X_{n2}, respectively. Each new node added on the graph is associated in the code to its corresponding image, regardless the trajectory it belongs to. In this way, with the node ID one can find it associated image, and with an image identifier, one can find its associated node ID.

4. Search for inter-session loop closings between the last image of session 2 and all images of session 1 using the algorithm explained in Section 2.2.3. Those candidates of session 1 retrieved by HALOC that present a transform after the RANSAC discrimination process with several *inliers* lower than a pre-fixed parameter ($MinRansacInliers$), are discarded and considered false positives that can harm the result of the graph optimization. The rest are accumulated and considered true positives. Let us name the number of true positives that close a loop with the last image of trajectory 2 as N_{TP}. For each true positive, the system stores the next data: (a) The name of both images that close the loop, (b) the identifiers IDa and IDb of both nodes involved in the loop closing and (c) the transform between both images (Z_{IDbi}^{IDai}).
5. Let us denote the number of accumulated inter-session loop closings as N_{ALC}, initialized to 0 when both sessions are jointed. Then, $N_{ALC} = N_{ALC} + N_{TP}$. When $N_{ALC} = NIsLoopClosings$, where $NIsLoopClosings$ is preset at the beginning of the process, then the graph is optimized with all the new pose constraints, following the next steps:
    (a) Recover the node IDs of the images associated with each inter-session loop closing classified as true positive, and every corresponding transform.
    (b) Add one additional link in the graph between nodes $IDai$ and $IDbi$, which content is $Z_{IDbi}^{IDai}, \forall i, 1 \leqslant i \leqslant N_{ALC}$.
6. Run the graph bundle adjustment using the Levenberg Marquardt algorithm. Even if after a certain number of iterations no inter-session loop closings are found, the graph will be optimized as well, just to re-adjust the odometric trajectory estimates.
7. $N_{ALC} = 0$, $N_{TP} = 0$, $n1 = n1 + 1$ and $n2 = n2 + 1$.
8. Return to the first step, and iterate the process until both trajectories are finished. If one of the two trajectories finishes before the other one, the system keeps adding the corresponding nodes of the session that is still on course. Obviously, no additional inter-session loop closings will be found in this case, so every graph optimization will include only the pose estimates given by the visual odometry of the ongoing mission.

The idea is illustrated by Algorithm 2.

Algorithm 2: Multi-robot Visual Graph SLAM.

Inputs

X_{n1}, X_{n2}: Last odometric displacements of Trajectories 1 and 2 before the map joining.

N_{n1}, N_{n2}: Identifiers (ID) of the last graph nodes corresponding to trajectories 1 and 2, after the map joining.

X^{n1},X^{n2}: Global poses corresponding to nodes N_{n1} and N_{n2}, after the map joining. $X^{n1} = X^{n1-1} \oplus X_{n1}$

I_{n1}, I_{n2}: last images taken by robots 1 and 2 at instants $n1$ and $n2$, just before the map joining.

Parameters

MinRansacInliers: Minimum number of RANSAC inliers to consider a transform between two images as a true positive

NIsLoopClosings: Maximum number of accumulated inter-session loop closings.

N_{min}: Minimum number of correspondences to consider a roto-translation as candidate.

N_c: Number of image candidates to be searched in Trajectory 1 to close a loop with I_{n2}

Variables

I_{1j}: Jth image of the first session, candidate to close a loop with a query image, found using HALOC.

N_{TP}: Number of images of trajectory 1 considered as true positives that close a loop with I_{n2}

N_{ALC}: Number of accumulated inter-session loop closings.

IDa_n, IDb_n, Ia_n, Ib_n: Graph nodes involved in the nth inter-session loop closing and images corresponding to each node.

$Z^{IDa_n}_{IDb_n}$: Transform (x, y, θ) associated with the nth inter-session loop closing.

ListOfCandidates: Structure that contains the list of image candidates to close a loop with a given query. Every element of the structure stores the image Id, its HALOC hash, and the number of features.

$H_{trajectory1}$: List of Hashes (global descriptor) type HALOC of all images of trajectory 1.

Functions

$[X] = RansacEstimateMotion(I_1, I_2)$: returns the odometric displacement (X) between images I_1 and I_2 using Algorithm 1,

$H = hash(I)$: is the function of the HALOC library that returns the HALOC global descriptor H of image I

$[ListOfCandidates] = LibHALOC(H_{trajectory1}, N_c, H_{I2})$: is the function of the HALOC library that gets N_c candidates of the trajectory 1 to close a loop with the query I_2.

AddRelativePose(Z^1_2, I_1, I_2): adds a new pose constraint (link with transform $Z^{I_1}_{I_2}$) between two graph nodes, I_1 and I_2

OptimizeGraph(): Does the global bundle adjustment of the whole graph using the Levenberg-Marquard algorithm.

begin
    $N_{ALC} = N_{TP} = n = 0$;
    Robot 1 takes the next image $\rightarrow In1 + 1$;
    Robot 2 takes the next image $\rightarrow In2 + 1$;
    H1=hash($In1 + 1$); H2=hash($In2 + 1$) ;
    $H_{trajectory1} \leftarrow$ H1 ;
    $[X_{n1+1}]$ = RansacEstimateMotion($In1, In1 + 1$).;
    $[X_{n2+1}]$ = RansacEstimateMotion($In2, In2 + 1$).;
    $X^{n1+1} = X^{n1} \oplus X_{n1+1}$. $N_{n1+1} \rightarrow$ node graph ID of X^{n1+1} ;
    $X^{n2+1} = X^{n2} \oplus X_{n2+1}$. $N_{n2+1} \rightarrow$ node graph ID of X^{n2+1} ;
    Store the correspondences $N_{n1+1} \rightarrow In1 + 1$ and $N_{n2+1} \rightarrow In2 + 1$; ;
    $[ListOfCandidates] = LibHALOC(H_{trajectory1}, N_c, H2)$;
    for $j \leftarrow 0$ **to** N_c **do**
        store $Ia_n = I_{1j}$, $Ib_n = I_{n2+1}$;
        $[Z^{IDa_n}_{IDb_n}] = RansacEstimateMotion(Ia_n, Ib_n)$;
        if *Number of Inliers between* Ia_n *and* $Ib_n \geqslant MinRansacInliers$ **then**
            $N_{TP} = N_{TP} + 1$;
            store $Z^{IDa_n}_{IDb_n}$ and IDa_n, IDb_n ;
            $n = n + 1$;
        end
    end
    N_{ALC}=N_{ALC}+N_{TP} ;
    $N_{TP} = 0$;
    if N_{ALC}=$NIsLoopClosings$ **then**
        for $i \leftarrow (n - N_{ALC})$ **to** $n + N_{ALC}$ **do**
            AddRelativePose ($Z^{IDa_i}_{IDb_i}$, IDa_i, IDb_i) ;
        end
        $N_{ALC} = 0$,
    end
    OptimizeGraph() ;
    $n1 = n1 + 1$, $n2 = n2 + 1$;
end

3. Experimental Results

3.1. Experimental Setup

A set of preliminary experiments were performed simulating the Multi-robot configuration with real underwater data. Three different datasets were used. Every dataset consists of two different video sequences, partially overlapping. Sequences were recorded either by a diver or by an AUV, both carrying a bottom looking camera with its lens axis perpendicular to the longitudinal axis of the vehicle or of the diver. Divers or autonomous underwater vehicles moved at an approximate constant altitude with respect to the sea bottom. The lack of any other sensorial data which could be supplied by an AUV, makes the localization system a pure vision-based approach. The three datasets have been recorded in several coastal sites of the north and south of Mallorca, at depths between 5 and 13 m. All the environments where the datasets were grabbed present a great variety of bottom textures, including seagrass, stones, sand, algae, moss and pebbles. The first dataset was recorded in the north coast of the island, by a diver with an attached Gopro camera, pointing downwards, moving on the surface at an approximate constant altitude of 4 m. Let us refer to both video sequences of this first dataset as $S11$ and $S12$. The camera altitude was obtained at the beginning of the video sequence by means of a visual marker of known size, placed at the sea bottom, in the starting point of each trajectory.

A second dataset formed by two partially overlapping trajectories named $S21$ and $S22$ were recorded also in the north coast of Mallorca, also by a diver supplied with a Gopro, looking to the bottom, far from $S11$ and $S12$, swimming on the water surface, at an approximate constant altitude of 4 m. In this case, the initial altitude was computed thanks to the know dimensions of a structure formed by markers and PVC tubes placed at the sea floor in the origin of both trajectories. The video resolution was 1920×1080 pixels, grabbed at 30 frames per second (fps), and prior to their use, all images were scaled down to 320×180 pixels.

A third dataset, with two video sequences named $S31$ and $S32$, was recorded by a SPARUS II AUV [52] property of the University of the Balearic Islands, at 7.5 fps, moving at a constant altitude of 3 m, in an area of the south of the island with an almost constant depth of 16 m. The navigation altitude is obtained from the vehicle navigation filter which integrates a Doppler Velocity Log (DVL), an Inertial Measurement Unit (IMU), a pressure sensor, an Ultra Short Baseline (*USBL*) acoustic modem [53], and a stereo 3D odometer. This dataset permitted to test the approach in larger environments with complex imagery due to the presence of sea grass on the sea bottom. In particular, $S31$ was recorded during a trajectory of 93 m long, and $S32$ during a a trajectory of 114 m, covering both an approximate area of 300 M^2 each one.

Figure 4 shows some samples of images included in the three datasets.

All these images show how all regions are colonized with *Posidonia oceanica*, a seagrass that forms dense and large meadows. Images of dataset 3 show a lack of illumination which increases at larger depths. With these conditions, the feature matching process decreases its performance and affects directly the accuracy of the visual odometry and the loop closing detection using HALOC. In this type of marine environments and with our robot and its equipment, moving at approximately 1 knot at altitudes between 3 m and 5 m, in areas with a depth between 16 m and 20 m, gave a good tradeoff between image overlap and illumination conditions.

Due to the particular texture of the Posidonia and the slight motion of its leafs caused by the currents, tracking stable visual features in consecutive overlapping frames is complicated and requires an accurate selection of the type of features and the feature detection/tracking parameters. Errors in this process will compromise the accuracy of the visual odometry and the image registration task for the loop closing confirmation. Previous pieces of work [22,41,42] already showed the high efficiency of SIFT features for underwater SLAM in areas colonized with Posidonia, in all the tasks involved in the process: visual odometry, image hashing with HALOC, loop closing detection, and pose refinement. Although SIFT feature detector is slower than other descriptors and delays the RANSAC-based matching process of Algorithm 1, given the robustness and traceability of SIFT, and according to our experience, this additional processing time is preferable to obtain more reliable trajectories than using other simpler features that take less time than SIFT to be computed and tracked, but can cause larger inaccuracies in the camera trajectory estimation or in the image registration process.

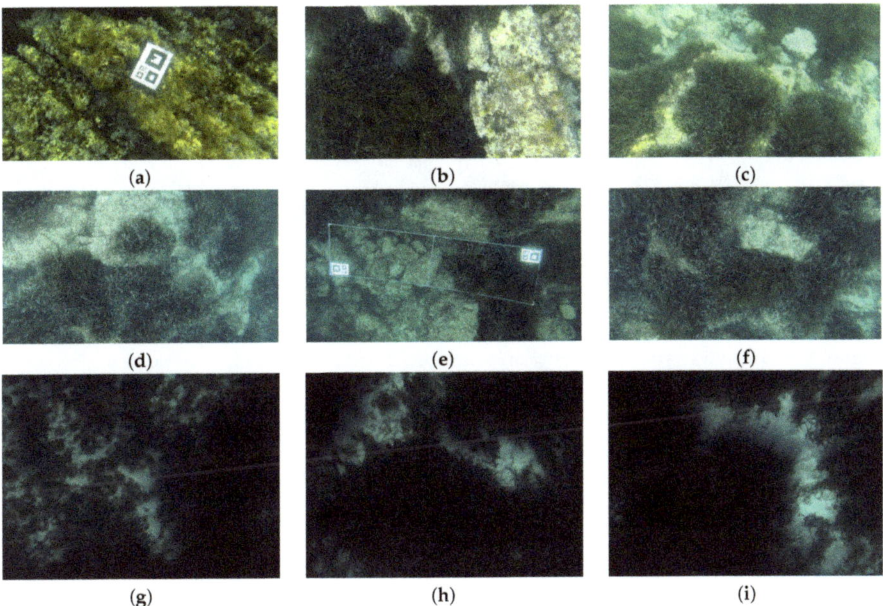

Figure 4. Examples of images of dataset 1, in (**a–c**), dataset 2, in (**d–f**), and dataset 3, in (**g–i**).

3.2. Experiments and Results

All experiments performed to get these first preliminary results were run offline, simulating the Multi-robot configuration with the visual data obtained in the sea. Key images of each video sequence that forms the different datasets mentioned in the previous section were extracted, indexed, stored separately in a hard disk, and processed consecutively according to the algorithms exposed in this paper.

Successive field experiments showed that, an overlap between consecutive images of 35% to 50% was necessary to obtain a robust visual odometry. On the other hand, reducing the number of images stored was also required in order to save as much memory space as possible. Consequently, a good trade off between both requirements was obtained selecting the keyframes of datasets 1 and 2 down-sampling the initial video frame rates at 1.1 fps, on average, and the dataset 3 at 3 fps. 226 key images were extracted from $S11$ and 199 from $S12$. 152 key frames were extracted from the video sequence $S21$ and 57 from $S22$. Finally, a total of 400 key images were extracted from dataset 3, 200 belonging to $S31$ and 200 to $S32$.

Local SLAMs are continuously executed for both trajectories in sequential steps of a predefined number of frames N; each local trajectory is accumulated and optimized from N to N frames, alternating both sessions every N frames. N is set differently for each dataset.

Figure 5 shows several samples of inter-session loop closings found by HALOC and confirmed by Algorithm 1.

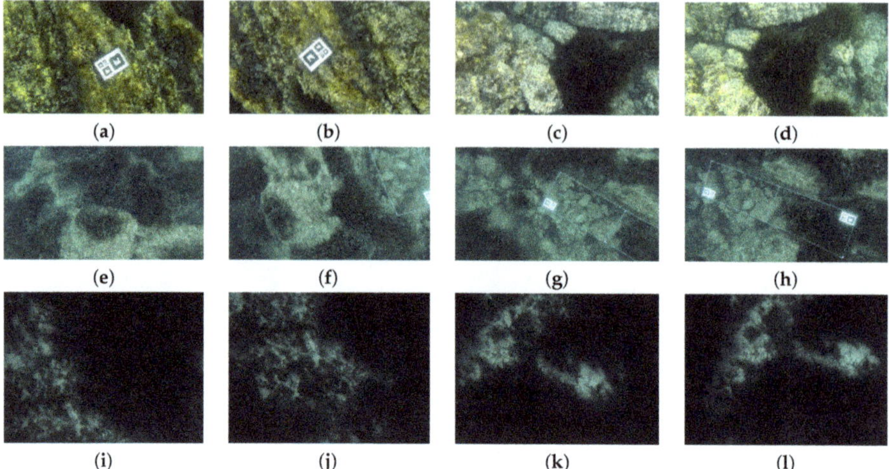

Figure 5. (**a–d**) Multi-session loops between sequences $S11$ and $S12$. (**e–h**) Multi-session loops between sequences $S21$ and $S22$. (**i–l**) Multi-session loops between sequences $S31$ and $S32$. (**a**) Closes a loop with (**b**,**c**) closes a loop with (**d**,**e**) close a loop with (**f**,**g**) close a loop with (**h**,**i**) closes a loop with (**j**,**k**) closes a loop with (**l**).

Once images and inter-session loops are available, the whole localization and mapping process starts. In other words, the sequence of actions is as follows:

1. For each dataset, extract the key images of both video sequences and store them in separated folders
2. For each dataset, compute the HALOC global descriptor of each image extracted from both video sequences.
3. For each dataset, compute and store in a file, the odometry, frame to frame, for both stored image sets, corresponding to both sessions.
4. At this point, for each dataset and for each of their sessions, the key frames and the odometry have been stored and related through successive identifiers. Thereafter, for each dataset run the local SLAM procedure, which:
    (a) Starts algorithm of Section 2.2, building the state vector of each session, by steps of N consecutive frames, using the displacements included in each odometry file.
    (b) For each newly gathered image (lets call it, the query image), searches for local loop closings on other images of the same dataset which positions are near the query. This search is done only among the images gathered before the query.
    (c) Optimize both local graphs according to Section 2.2.2.
    (d) For each image of trajectory 2 (called *the query*), the algorithm searches the best 5 HALOC loop closing potential candidates of trajectory 1. Each candidate, if any, is confirmed by means of Algorithm 1, and filtered out if the number of inliers is lower than the predefined threshold.
    (e) Accumulate the number of inter-session true loop closings.
    (f) When the number of accumulated inter-session loop closings is greater than a certain threshold, join both sessions in a single pose-based graph. That means transforming all members of the joined state vector in global poses and the corresponding graph nodes, associating to each node the corresponding image.
5. Run the Multi-robot SLAM procedure, according to Algorithm 2

(a) Obtain the rest of images from memory and add new nodes according to the successive odometry data of both sessions, as explained in Section 2.3.
(b) Search among all images of session 1 the best 5 candidates to close an inter-session loop with each new query of session 2, and filter out all those that do not present enough inliers after running Algorithm 1. HALOC is, obviously, the method used to find these candidates to close loops inter-sessions. Since each image will be associated with a node of the global graph, computing the transform between two candidates to close a loop and adding this transform between both nodes will be straightforward.
(c) Get the transform between pairs of images that constitute true positives (true loop closings).
(d) Add this transform to the graph as a new pose constrain, in the form of links between two nodes. The nodes will be those related with the images involved in the inter-session loop closing.
(e) Optimize the graph.
(f) Finish the process when all images from both sessions have been already used.

This simulation does not permit assessing anything related to execution times because, although the whole process works with real data, the Multi-robot configuration has been simulated, split into three different software packages of different natures. The purpose of these preliminary results is not giving an accurate set of quantitative and numerical results to assess the process in terms of execution time or trajectory accuracy. The aim of this section is presenting the implementation of a new approach and a set of preliminary results that provide: (a) a proof of viability and feasibility of the solution, (b) a proof of its utility and suitability to manage, in a single map, two sessions of two different robots that operate simultaneously, in a simple way, (c) a qualitative proof of concept and, (d) the source code and a dataset to be tested, open to further improvements.

Obtaining a ground truth trajectory underwater is a challenging task, unless one can install an infrastructure of acoustic beacons or Long Baseline (LBL) systems, which is costly and complicated to run and manage, and imposes spatial restrictions on the motion of the robots. In our experiments, there is no ground truth and no possibility to get it. The planned trajectories, in the case of those performed with the AUV, cannot be used as a ground truth either, since they differ substantially from the ones that the AUV ends up performing (which is usual in underwater robotics). In consequence, we cannot compare the trajectories estimated by the system with another one that serves as reference. In our case, robustness has been qualitatively validated by two means, (a) comparing the resulted global maps with the mosaics obtained with BIMOS [54] and, (b) comparing the direct transforms between images that close loops, obtained with Algorithm 1 with the transforms between the graph-nodes related to the same loop closing images. These two points have been already used and validated in previous pieces of work [46].

Figure 6a,b,d,e show, respectively, the trajectory of $S11$, $S12$, $S21$ and $S22$ estimated by the local SLAM procedure described in Section 2.2. Figure 6c,f show, respectively, the global graph obtained applying the Multi-robot SLAM procedure described in Section 2.3.

Figures 7 and 8 show four photo-mosaics corresponding to sequences $S11$, $S12$, $S21$ and $S22$. These photomosaics have been obtained using BIMOS [54], a mosaicing algorithm based on bags of binary words that already demonstrated a great performance in underwater environments [55,56].

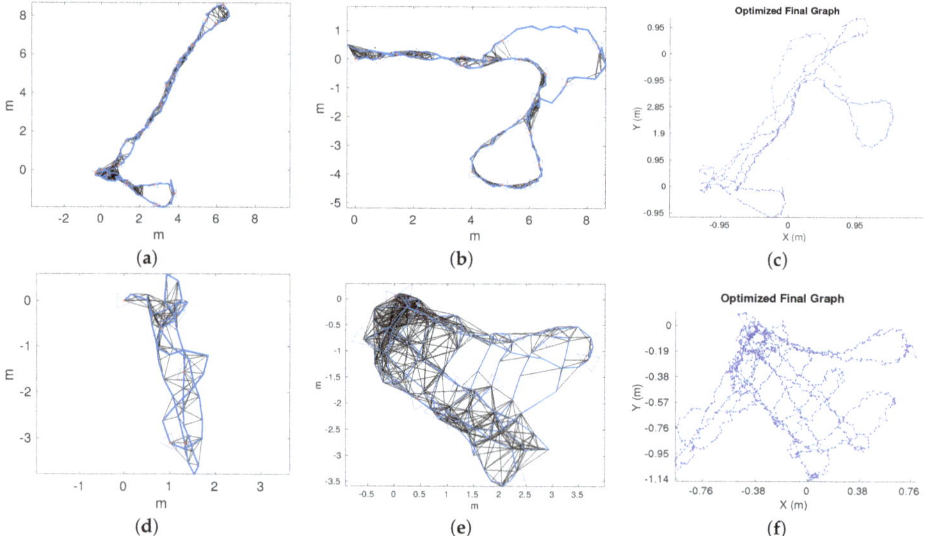

Figure 6. (**a**) Trajectory S11, (**b**) Trajectory S12, (**c**) Multi-robot final graph from S11 and S12. (**d**) Trajectory S21, (**e**) Trajectory S22, (**f**) Multi-robot final graph from S21 and S22.

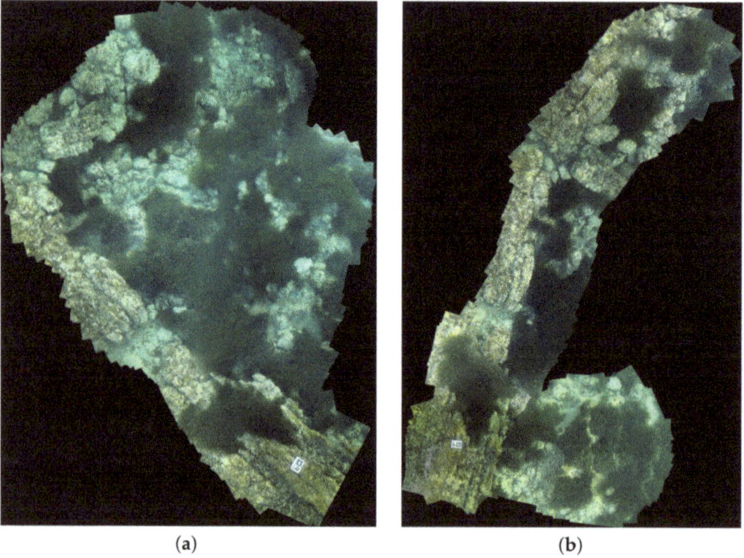

Figure 7. (**a**) Photo-mosaic of S11. (**b**) Photo-mosaic of S12.

The resulting mosaics have associated an implicit trajectory which imposes the position of each image with respect to the origin of the mosaic system of coordinates. Due to the lack of any trajectory ground truth and the impossibility to get it, the mosaic is, to a certain extent, a qualitative reference to assess the quality of the resulting joined trajectories of Figure 6c,f, since BIMOS has already demonstrated its good performance in land and underwater. Notice how the mosaic of Figure 7b shows a montage very close to the SLAM trajectory of S11, and Figure 7a a mosaic fitting the SLAM trajectory of $S12$. It can be assumed that the quality of the single and joined graphs obtained with BIMOS are similar to the quality of the mosaics.

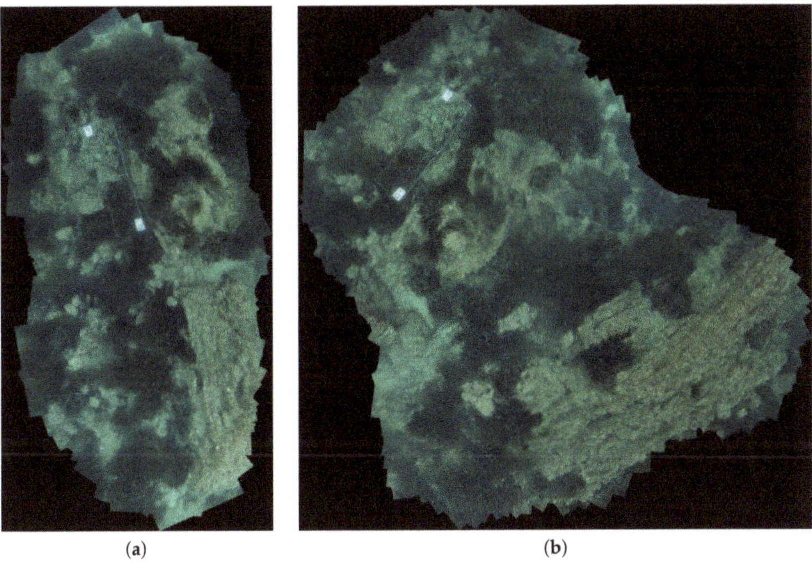

(a) (b)

Figure 8. (a) Photo-mosaic of $S21$. (b) Photo-mosaic of $S22$.

If both mosaics are aligned by their left laterals, where the marker is located, something that fits very well with the global graph is obtained. The same applies to the mosaics of Figure 8. The structure containing the two markers is the point joining both trajectories. Then, the alignment of both figures by the markers gives a result that also fits perfectly with the global graph of Figure 6.

Figure 9a,b show the local SLAM trajectory of $S31$ and $S32$, and Figure 9c shows the global graph estimated after joining both trajectories and applying the Multi-robot graph SLAM approach.

An illustrative video of the whole process involving the three datasets can be seen in [57]. The video shows, at the beginning, some sequences grabbed underwater and used to test our approach. Afterwards, it shows the whole process for the three datasets exposed: (1) The local SLAM for both separated trajectories, (2) the moment when both sessions are joined and converted into a single global graph, and (3) how the graph continues growing in different directions, each one corresponding to each trajectory involved in the Multi-robot mission. As mentioned in previous sections, the joined graph is optimized every time a set of inter-session loop closings are confirmed.

As this is a pure visual-based SLAM approach and no other sensorial input is included in the multi-localization process, this is firstly computed in pixel units. To find the relation between pixels and metric units, the known real dimensions of the markers used to establish the starting and end points of each trajectory (see pictures (a) and (e) of Figure 4) were related with the pixel dimensions of the markers in the images. These relations resulted in coefficients ranging between 0.0015 and 0.0019, depending on each video sequence.

Another way to verify the consistency of the resulting optimized global map is comparing the transform between two images (called I1 and I2) that close an inter-session loop, when it is obtained by two different means: (1) using Algorithm 1, and (2) running the next operation: $\ominus P1 \oplus P2$, where $P1$ is the global pose associated with the graph node corresponding to I1, and $P2$ is the global pose associated with the graph node corresponding to I2. The idea is illustrated in Figure 10: $P1$ and $P2$ are the global poses associated with two loop closing nodes, T_{I2}^{I1} is the direct transform between P1 and P2 obtained with the RANSAC-based algorithm applied directly on I1 and I2. In principle, if the graph is consistent, T_{I2}^{I1} has to be equivalent to the inverse of $P1$ composed with $P2$, which is the transform between $I1$ and $I2$, but obtained composing the global poses between the respective nodes of the loop closing images.

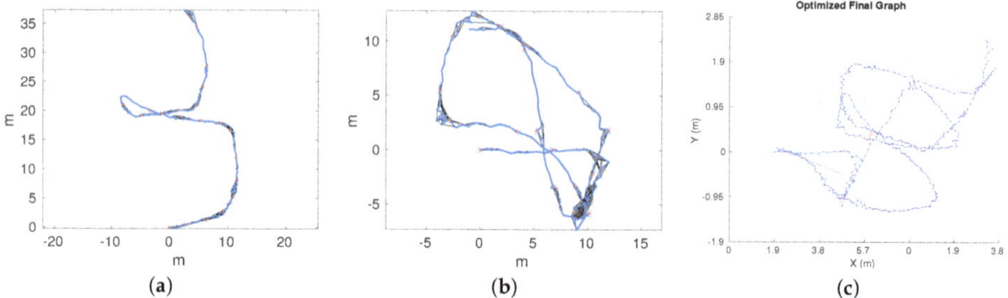

Figure 9. (**a**) Trajectory $S31$, (**b**) Trajectory $S32$, (**c**) Multi-robot final graph from $S31$ and $S32$.

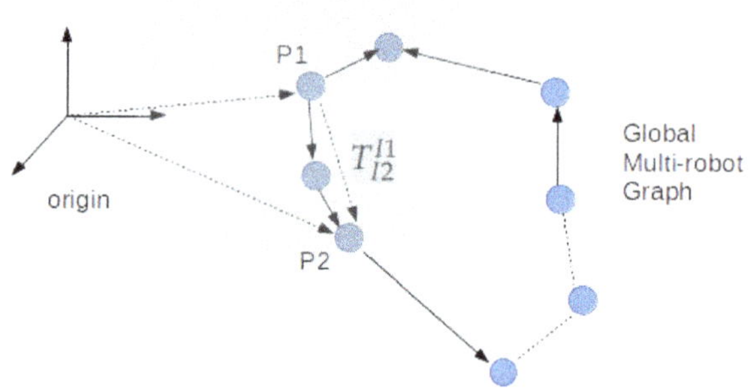

Figure 10. Transforms between nodes P1 and P2.

Tables 1 and 2 show some samples of quantitative results of intersession loop closings with the corresponding transforms calculated by both aforementioned ways. The first to forth columns show, respectively, the number of both images that close an inter-session loop and the graph nodes which each image is associated with. Column I1 contains images of the sessions $S11$ and $S21$, and column I2 contains images of sessions $S12$ and $S22$. Fifth and sixth columns indicate the 2D transform, in translation and rotation (x, y, θ), computed indirectly through the graph and directly using RANSAC (Algorithm 1). The units of these transforms are expressed in pixels and radians. The seventh column indicates the difference between both transforms, in module (meters) and orientation (radians). These samples indicate that: (a) for $S11$ and $S12$, the difference of transforms ranges between 1.8 cm and 0.15 mm in module and between 0.05 rad. (2.86°) and 0.0089 rad. (0.5°) in orientation, and (b) the difference for $S21$ and $S22$ ranges between 1.02 cm and 14.8 cm in module and between 0.0085 rad. (0.5°) and 0.21 rad. (12.03°) in orientation. These differences are totally acceptable, taking into account that there are errors inherent to the RANSAC transform estimation process due to the possible (and usual) presence of any inconsistent inlier, and differences (or errors) due to the successive graph optimizations, which also cause subsequent readjustments of all node poses.

Table 1. Comparison of Transforms between images of $S11$ and $S12$.

I1	I2	Node I1	Node I2	Graph Transform	Ransac Transform	dif:(mod., yaw)
56	280	115	113	[3.04;−4.05;0.1711]	[3.12;−3.99;0.18]	(0.00015 m, 0.0089 rad)
64	287	131	126	[12.10;41.17;0.0633]	[21.71;33.31;−0.03]	(0,018 m, 0.01 rad)
152	420	307	393	[37.09;31.95;−0.34]	[36.65;31.47;−0.29]	(0,00097 m, 0.05 rad)

Table 2. Comparison of Transforms between images of $S21$ and $S22$.

I1	I2	Node I1	Node I2	Graph Transform	Ransac Transform	dif:(mod., yaw)
335	1181	672	602	[30.43;59.40;2.42]	[−18.61;−1.69;2.61]	(0.148 m, 0.2 rad)
603	1190	941	621	[1.19;−30.22;−0.024]	[11.77;19.69;0.1959]	(0.097 m, 0.21 rad)
877	1211	1215	662	[−6.34;−92.66;−0.196]	[24.69;−50.403;−0.314]	(0.099 m, 0.12 rad)
871	1211	1209	662	[−20.65;−84.08;-3.69×10^{-4}]	[14.75;−43.66;−0.086]	(0.0102 m, 0.085 rad)

3.3. Some Considerations of the Data Reduction

The length of the HALOC global descriptors used in the aforementioned tests is 384 floats. That is, a total of 1536 bytes per image, considering that in C + + a float needs 4 bytes for memory storage. All this, regardless the image resolution and the amount of SIFT features per image. That means that no matter how big is the image and how many visual features are being detected per image, that the length of the hash maintains invariable. Conversely, the size of a color image with a very reduced resolution of 320×240 pixels would be $320 \times 240 \times 3$ = 230,400 bytes. The save on memory space for data storage is clearly reduced when using the HALOC hash instead of the original images. The set of image features must be stored for every image, in any case, since they are needed in the later processes of loop closing confirmation. However, the computational cost of comparing two hashes to retrieve the best candidates for loop closing just calculating the L1-norm of two vectors is much lower than finding the best candidates with a brute-force recursive feature double-matching with RANSAC. From this point on, one can think of applying additional strategies to limit the communication between robots, complementing, for instance the solution proposed in [35], where the images are sent only among robots that view, simultaneously, a common point. In this case, we introduce an additional layer to compress the information to be exchanged, since instead of sending JPEG images, robots would send their respective hashes.

3.4. Sources Availability

The source code for the odometry computation has been developed in Matlab and it is available at [58]. The source code of the HALOC library is available at [59], for its C + + version, and in [60], for its version in Python.

The sources for the local SLAM, the Map Joining and the later Multi-Robot Graph SLAM have been developed also in Matlab, and they are available for the community at [61]. The pose-based graph management has been programmed using the Matlab library for localization and pose estimation especially addressed to mobile autonomous vehicles [62].

4. Discussion, Conclusions and Future Work

This paper presents a new approach to visual SLAM for Multi-robot configurations, based on joining, in a single pose-based graph, several trajectories of different robots which operate simultaneously in a common area of interests. The system finds loop closings between images of different robot trajectories by means of a hash-based methodology (HALOC), and uses them to add additional constraints to the global graph. As exposed in the text, the use of HALOC clearly guarantees an important reduction in storage space, amounts of data to be transferred and time dedicated for loop closing detection, especially in centralized multi-robot configurations. Using HALOC also assures a proper graph optimization since it has already been proved underwater, showing excellent results in terms of success ratios in loop closing detection.

The strategy for map joining comes from [46], but adapted to a Multi-robot configuration, which differs from a multi-session case in some aspects. This strategy is simple, easy to replicate, effective, and, more importantly, as flexible as possible to modulate the moment for map joining depending on the mission conditions and convenience, which implies a trade off between the accuracy of the local maps before they are joined, or the need for joining as soon as possible all the trajectories to centralize the global multiple localization of all the robotic team in a single agent.

Preliminary experiments permitted to show how the application of the new approach for joining, online, multiple ongoing sessions was perfectly feasible, suggesting a certain consistency and reliability in the results, from a qualitative point of view.

Although, until now, we focused our efforts exclusively in the estimation of the camera pose and trajectory, this algorithm has been designed to be applied on board a vehicle. Therefore, one priority ongoing task is testing this algorithm in a team of real vehicles operating in the sea. To this end, a ROS [63] wrapper in C++ is currently being developed and tested.

We have now focused our efforts exclusively in the estimation of the camera pose and trajectory assuming that navigation and control are solved issues. In fact, most of the existing research on SLAM makes the same assumption. However, the continuous re-estimation of the vehicle poses thanks to the SLAM algorithm surely affects the control of the vehicles, because the control modules are fed with the poses and velocities. In addition, changes in control affect, in turn, the vehicle navigation. At the moment the SLAM algorithm is completely decoupled from the control module, but once it is installed on a vehicle, the vehicles velocity and pose provided by our SLAM modules, together with the mission goal points, will be input in the navigation and control subsystems. Another possible line of research that is also under consideration is to make the goal points also depend on SLAM in order to add exploration to the AUV capabilities.

Other future work includes:

(1) Extending the tests to additional environments with longer trajectories. (2) Extending the assessment of the approach by means of evaluating the performance of the SLAM pose corrections in the presence of additive Gaussian noise in the visual odometry, and all evaluation techniques employed in [46]. (3) Comparing with other Multi-robot software packages still not tested underwater, such as DSLAM.

Matlab sources are available in a public repository giving the chance to the scientific community of testing, replicating, and also improving them.

Author Contributions: Both authors contributed equally to this work, including the Conceptualization, theoretical methodology, the implementation of the software, the validation with the datasets obtained in the sea, writing the original draft, and the final supervision. All authors have read and agreed to the published version of the manuscript.

Funding: This work is partially supported by Ministry of Economy and Competitiveness under contract DPI2017-86372-C3-3-R (AEI,FEDER,UE).

Conflicts of Interest: The authors declare no conflict of interest.

References

1. Durrant-Whyte, H.; Bailey, T. Simultaneous Localization and Mapping (SLAM): Part I The Essential Algorithms. *Robot. Autom. Mag.* **2006**, *2*, 99–110. [CrossRef]
2. Burguera, A.; González, Y.; Oliver, G. Underwater SLAM with Robocentric Trajectory Using a Mechanically Scanned Imaging Sonar. In Proceedings of the International Conference on Intelligent Robotis and Systems (IROS), San Francisco, CA, USA, 25–30 September 2011; pp. 3577–3582.
3. Ferri, G.; Djapic, V. Adaptive Mission Planning for Cooperative Autonomous Maritime Vehicles. In Proceedings of the IEEE International Conference on Robotics and Automation (ICRA), Karlsruhe, Germany, 6–10 May 2013; pp. 5586–5592.

4. Conte, G.; Scaradozzi, D.; Mannocchi, D.; Raspa, P.; Panebianco, L.; Screpanti, L. Development and Experimental Tests of a ROS Multi-agent Structure for Autonomous Surface Vehicles. *J. Intell. Robot. Syst.* **2018**, *92*, 705–718. [CrossRef]
5. Davison, A.; Calway, A.; Mayol, W. Visual SLAM. *IEEE Trans. Robot.* **2007**, *24*, 1088–1093. [CrossRef]
6. McDonald, J.; Kaess, M.; Cadena, C.; Neira, J.; Leonard, J.J. Real-time 6-DOF Multi-session Visual SLAM over Large-Scale Environments. *Robot. Auton. Syst.* **2013**, *61*, 1144–1158. [CrossRef]
7. Abdulgalil, M.; Nasr, M.; Elalfy, M.; Khamis, A.; Karray, F. Multi-Robot SLAM: An Overview and Quantitative Evaluation of MRGS ROS Framework for MR-SLAM. In *International Conference on Robot Intelligence Technology and Applications*; Springer: Cham, Switzerland, 2019; Volume 751, pp. 165–183.
8. Mahdoui, N.; Frémont, V.; Natalizio, E. Communicating Multi-UAV System for Cooperative SLAM-based Exploration. *J. Intell. Robot. Syst.* **2019**, *46*, 1–19. [CrossRef]
9. Shkurti, F.; Chang, W.; Henderson, P.; Islam, M.; Gamboa, J.; Li, J.; Manderson, T.; Xu, A.; Dudek, G.; Sattar, J. Underwater Multi-robot Convoying Using Visual Tracking by Detection. In Proceedings of the International Conference on Intelligent Robots and Systems (IROS), Vancouver, BC, Canada, 24–28 September 2017; pp. 4189–4196.
10. Allotta, B.; Costanzi, R.; Ridolfi, A.; Colombo, C.; Bellavia, F.; Fanfani, M.; Pazzaglia, F.; Salvetti, O.; Moroni, D.; Pascali, M.A.; et al. The ARROWS Project: Adapting and Developing Robotics Technologies for Underwater Archaeology. *IFAC-PapersOnLine* **2015**, *48*, 194–199. [CrossRef]
11. Howard, A. Multi-Robot Simultaneous Localization and Mapping Using Particle Filters. *Int. J. Robot. Res.* **2006**, *25*, 1243–1256. [CrossRef]
12. Kaess, M.; Ranganathan, A.; Dellaert, F. iSAM: Incremental Smoothing and Mapping. *IEEE Trans. Robot. (TRO)* **2008**, *24*, 1365–1378. [CrossRef]
13. Kim, B.; Kaess, M.; Fletcher, L.; Leonard, J.; Bachrach, A.; Roy, N.; Teller, S. Multiple Relative Pose Graphs for Robust Coopetative Mapping. In Proceedings of the IEEE International Conference on Robotics and Automation (ICRA), Anchorage, AK, USA, 3–7 May 2010..
14. Schuster, M.J.; Schmid, K.; Brand, C.; Beetz, M. Distributed stereo vision-based 6D localization and mapping for multi-robot teams. *J. Field Robot.* **2019**, *36*, 305–332. [CrossRef]
15. Saeedi, S.; Trentini, M.; Seto, M.; Li, H. Multiple-Robot Simultaneous Localization and Mapping: A Review. *J. Field Robot.* **2016**, *33*, 3–46. [CrossRef]
16. Angeli, A.; Doncieux, S.; Meyer, J.; Filliat, D. Real-Time Visual Loop-Closure Detection. In Proceedings of the IEEE International Conference on Robotics and Automation (ICRA), Pasadena, CA, USA, 19–23 May 2008; pp. 1842 –1847.
17. Kim, A.; Eustice, R.M. Real-Time Visual SLAM for Autonomous Underwater Hull Inspection Using Visual Saliency. *IEEE Trans. Robot.* **2013**, *29*, 719–733. [CrossRef]
18. Monga, V.; Evans, B.L. Perceptual Image Hashing Via Feature Points: Performance Evaluation and Tradeoffs. *IEEE Trans. Image Process.* **2006**, *15*, 3452–3465. [CrossRef]
19. Liu, Y.; Zhang, H. Visual Loop Closure Detection with a Compact Image Descriptor. In Proceedings of the IEEE/RSJ International Conference on Intelligent Robots and Systems (IROS), Vilamoura, Portugal, 7–12 October 2012; pp. 1051–1056.
20. Arandjelovic, R.; Zisserman, A. All About VLAD. In Proceedings of the IEEE Conference on Computer Vision and Pattern Recognition (CVPR), Portland, OR, USA, 23–28 June 2013; pp. 1578–1585.
21. Bonin-Font, F.; Negre, P.; Burguera, A.; Oliver, G. LSH for Loop Closing Detection in Underwater Visual SLAM. In Proceedings of the 2014 IEEE Emerging Technology and Factory Automation (ETFA), Barcelona, Spain, 16–19 September 2014; pp. 1–4.
22. Negre Carrasco, P.L.; Bonin-Font, F.; Oliver-Codina, G. Global Image Signature for Visual Loop-closure Detection. *Autonom. Robots* **2016**, *40*, 1403–1417. [CrossRef]
23. Jain, U.; Namboodiri, V.; Pandey, G. Compact Environment-Invariant Codes for Robust Visual Place Recognition. In Proceedings of the 14th Conference on Computer and Robot Vision (CRV), Edmonton, AB, Canada, 17–19 May 2017; pp. 40–47.
24. Glover, A.; Maddern, W.; Warren, M.; Reid, S.; Milford, M.; Wyeth, G. Openfabmap: An Open Source Toolbox for Appearance-based Loop Closure Detection. In Proceedings of the IEEE International Conference on Robotics and Automation (ICRA), Saint Paul, MN, USA, 14–18 May 2011.

25. Cieslewski, T.; Choudhary, S.; Scaramuzza, D. Data-Efficient Decentralized Visual SLAM. In Proceedings of the IEEE International Conference on Robotics and Automation (ICRA), Brisbane, QLD, Australia, 21–25 May 2018; pp. 2466–2473.
26. Arandjelović, R.; Gronat, P.; Torii, A.; Pajdla, T.; Sivic, J. NetVLAD: CNN Architecture for Weakly Supervised Place Recognition. *IEEE Trans. Pattern Anal. Mach. Intell.* **2018**, *40*, 1437–1451. [CrossRef] [PubMed]
27. Mur-Artal, R.; Montiel, J.M.; Tardos, J.D. ORB-SLAM: A Versatile and Accurate Monocular SLAM System. *IEEE Trans. Robot.* **2015**, *31*, 1147–1163. [CrossRef]
28. Rublee, E.; Rabaud, V.; Konolige, K.; Bradski, G. ORB: An Efficient Alternative to SIFT or SURF. In Proceedings of the 2011 International Conference on Computer Vision (ICCV), Barcelona, Spain, 6–13 November 2011; pp. 2564–2571.
29. Zhang, P.; Wang, H.; Ding, B.; Shang, S. Cloud-Based Framework for Scalable and Real-Time Multi-Robot SLAM. In Proceedings of the 2018 IEEE International Conference on Web Services (ICWS), San Francisco, CA, USA, 2–7 July 2018; pp. 147–154.
30. Geiger, A.; Lenz, P.; Stiller, C.; Urtasun, R. The KITTI Vision Benchmark Suite. Available online: http://www.cvlibs.net/datasets/kitti/ (accessed on 7 May 2020).
31. Karrer, M.; Schmuck, P.; Chli, M. CVI-SLAM—Collaborative Visual-Inertial SLAM. *IEEE Robot. Autom. Lett.* **2018**, *3*, 2762–2769. [CrossRef]
32. Elibol, A.; Kim, J.; Gracias, N.; García, R. Efficient Image Mosaicing for Multi-robot Visual Underwater Mapping. *Pattern Recognit. Lett.* **2014**, *46*, 20–26. [CrossRef]
33. Young-Hoo, K.; Steven, L. Applicability of Localized-calibration Methods in Underwater Motion Analysis. In Proceedings of the XVIII International Symposium on Biomechanics in Sports; 2000. Available online: https://ojs.ub.uni-konstanz.de/cpa/article/view/2530 (accessed on 7 May 2020).
34. Hans-Gerd, M. *New Developments in Multimedia Photogrammetry*; Wichmann Verlag: Karlsruhe, Germany, 1995.
35. Pfingsthorn, M.; Birk, A.; Bülow, H. An Efficient Strategy for Data Exchange in Multi-robot Mapping Under Underwater Communication Constraints. In Proceedings of the IEEE/RSJ International Conference on Intelligent Robots and Systems, Taipei, Taiwan, 18–22 October 2010; pp. 4886–4893.
36. Paull, L.; Huang, G.; Seto, M.; Leonard, J. Communication-constrained Multi-AUV Cooperative SLAM. In Proceedings of the 2015 IEEE International Conference on Robotics and Automation (ICRA), Seattle, WA, USA, 26–30 May 2015; pp. 509–516.
37. Jaume I University of Castellón, IRSLab; University of Girona, CIRS; University of Balearic Islands. TWIN roBOTs for Cooperative Underwater Intervention Missions. 2019. Available online: http://www.irs.uji.es/twinbot/twinbot.html (accessed on 7 May 2020).
38. Burguera, A.; Bonin-Font, F.; Oliver, G. Trajectory-based visual localization in underwater surveying missions. *Sensors* **2015**, *15*, 1708–1735. [CrossRef]
39. Zhao, B.; Hu, T.; Zhang, D.; Shen, L.; Ma, Z.; Kong, W. 2D Monocular Visual Odometry Using Mobile-phone Sensors. In Proceedings of the 34th Chinese Control Conference (CCC), Hangzhou, China, 28–30 July 2015; pp. 5919–5924.
40. SRV Group. Systems, Robotics and Vision Group, University of the Balearic Islands. Available online: http://srv.uib.es/projects/ (accessed on 7 May 2020).
41. Negre Carrasco, P.L.; Bonin-Font, F.; Oliver, G. Cluster-based Loop Closing Detection for Underwater SLAM in Feature-poor Regions. In Proceedings of the IEEE International Conference on Robotics and Automation, Stockholm, Sweden, 16–21 May 2016; pp. 2589–2595. [CrossRef]
42. Peralta, G.; Bonin-Font, F.; Caiti, A. Real-time Hash-based Loop Closure Detection in Underwater Multi-Session Visual SLAM. In Proceedings of the Ocean 2019, Marseille, France, 17–20 June 2019.
43. Kümmerle, R.; Grisetti, G.; Strasdat, H.; Burgard, K.K.W. G2o: A General Framework for Graph Optimization. *ICRA. IEEE* **2011**, 3607–3613.
44. Newman, P.; Leonard, J.; Rikoski, R. Towards Constant-Time SLAM on an Autonomous Underwater Vehicle Using Synthetic Aperture Sonar. In Proceedings of the Eleventh International Symposium on Robotics Research, Sienna, Italy, 19–22 October 2003; Volume 15, pp. 409–420.
45. Muller, M.; Surmann, H.; Pervolz, K.; May, S. The Accuracy of 6D SLAM Using the AIS 3D Laser Scanner. In Proceedings of the 2006 IEEE International Conference on Multisensor Fusion and Integration for Intelligent Systems, Heidelberg, Germany, 3–6 September 2006; pp. 389–394.

46. Burguera, A.B.; Bonin-Font, F. A Trajectory-Based Approach to Multi-Session Underwater Visual SLAM Using Global Image Signatures. *J. Mar. Sci. Eng.* **2019**, *7*, 278. [CrossRef]
47. Smith, R.; Cheeseman, P.; Self, M. A Stochastic Map for Uncertain Spatial Relationships. In Proceedings of the 4th International Symposium on Robotics Research, Siena, Italy, 19–22 October 1987; pp. 467–474.
48. Kanzow, C.; Yamashita, N.; Fukushima, M. Levenberg-Marquardt Methods for Constrained Nonlinear Equations with Strong Local Convergence Properties. *J. Comput. Appl. Math.* **2002**, *172*, 375–397. [CrossRef]
49. Gavin, H. The Levenberg-Marquardt Method for Nonlinear Least Squares Curve-Fitting Problems. 2019. Available online: http://people.duke.edu/~hpgavin/ce281/lm.pdf (accessed on 10 May 2002).
50. Smith, R.; Self, M.; Cheeseman, P. A Stochastic Map for Uncertain Spatial Relationships. *Comput. Sci.* **1988**. Available online: https://www.semanticscholar.org/paper/A-stochastic-map-for-uncertain-spatial-Smith-Self/76a6c5352a0fbc3fec5395f1501b58bd6566d214 (accessed on 5 May 2002)
51. Grisetti, G.; Kuemmerle, R.; Stachniss, C.; Burgard, W. A Tutorial on Graph-Based SLAM. *Intell. Transp. Syst. Mag. IEEE* **2010**, *2*, 31–43. [CrossRef]
52. Carreras, M.; Hernandez, J.; Vidal, E.; Palomeras, N.; Ribas, D.; Ridao, P. Sparus II AUV—A Hovering Vehicle for Seabed Inspection. *IEEE J. Ocean. Eng.* **2018**, *43*, 344–355. [CrossRef]
53. Guerrero, E.; Bonin-Font, F.; Negre, P.L.; Massot, M.; Oliver, G. USBL Integration and Assessment in a Multisensor Navigation Approach for AUVs. *IFAC-PapersOnLine* **2017**, *50*, 7905–7910.
54. García, E.; Ortiz, A.; Bonnín, F.; Company, J.P. Fast Image Mosaicing using Incremental Bags of Binary Words. In Proceedings of the IEEE International Conference on Robotics and Automation (ICRA), Stockholm, Sweden, 16–21 May 2016.
55. Bonin-Font, F.; Massot, M.; Codina, G.O. Towards Visual Detection, Mapping and Quantification of Posidonia Oceanica using a Lightweight AUV. *IFAC-PapersOnLine* **2016**, 49, 500–505. [CrossRef]
56. García, E.; Ortiz, A. Hierarchical Place Recognition for Topological Mapping. *IEEE Trans. Robot.* **2017**, *33*, 1061–1074. [CrossRef]
57. Bonin-Font, F. Multi-Robot Visual Graph SLAM–An Ilustrative Video. Available online: https://www.youtube.com/watch?v=c7faAm2HIpc (accessed on 16 May 2020).
58. Burguera, A. Visual Odometry Sources. 2015. Available online: https://github.com/aburguera/VISUAL_ODOMETRY_2D (accessed on 16 May 2020).
59. Negre Carrasco, P.L. ROS C++ Library for Hash-Based Loop Closure (HALOC). Available online: https://github.com/srv/libhaloc (accessed on 2 June 2015).
60. Bonin-Font, F. Python Library for Hash-Based Loop Closure (HALOC). Available online: https://github.com/srv/HALOC-Python (accessed on 27 March 2019).
61. Bonin-Font, F.; Burguera, A. Sources of Multi-Robot Visual Graph SLAM with a Sample Dataset. Available online: https://github.com/srv/Multi-Robot-Visual-Graph-SLAM (accessed on 7 May 2020).
62. Matlab. Localization and Pose Estimation: Pose Graph. Available online: https://es.mathworks.com/help/nav/ref/posegraph.html (accessed on 1 January 2019).
63. Quigley, M.; Conley, K.; Gerkey, B.P.; Faust, J.; Foote, T.; Leibs, J.; Wheeler, R.; Ng, A.Y. ROS: An Open-Source Robot Operating System. In *ICRA Workshop on Open Source Software*; 2009. Available online: https://blog.acolyer.org/2015/11/02/ros-an-open-source-robot-operating-system/ (accessed on 1 May 2020).

Journal of
Marine Science and Engineering

MDPI

Article

Underwater Pipe and Valve 3D Recognition Using Deep Learning Segmentation

Miguel Martin-Abadal *, Manuel Piñar-Molina, Antoni Martorell-Torres, Gabriel Oliver-Codina and Yolanda Gonzalez-Cid

Departament de Matemàtiques i Informàtica, Universitat de les Illes Balears, Carretera de Valldemossa Km. 7.5, 07122 Palma, Spain; manuel.pinar@uib.es (M.P.-M.); antonimartorelltorres@gmail.com (A.M.-T.); goliver@uib.es (G.O.-C.); yolanda.gonzalez@uib.es (Y.G.-C.)
* Correspondence: miguel.martin@uib.es

Abstract: During the past few decades, the need to intervene in underwater scenarios has grown due to the increasing necessity to perform tasks like underwater infrastructure inspection and maintenance or archaeology and geology exploration. In the last few years, the usage of Autonomous Underwater Vehicles (AUVs) has eased the workload and risks of such interventions. To automate these tasks, the AUVs have to gather the information of their surroundings, interpret it and make decisions based on it. The two main perception modalities used at close range are laser and video. In this paper, we propose the usage of a deep neural network to recognise pipes and valves in multiple underwater scenarios, using 3D RGB point cloud information provided by a stereo camera. We generate a diverse and rich dataset for the network training and testing, assessing the effect of a broad selection of hyperparameters and values. Results show *F1-scores* of up to 97.2% for a test set containing images with similar characteristics to the training set and up to 89.3% for a secondary test set containing images taken at different environments and with distinct characteristics from the training set. This work demonstrates the validity and robust training of the PointNet neural in underwater scenarios and its applicability for AUV intervention tasks.

Keywords: point cloud segmentation; deep learning; pipe and valve recognition; underwater perception; computer vision

Citation: Martin-Abadal, M.; Piñar-Molina, M.; Martorell-Torres, A.; Oliver-Codina, G.; Gonzalez-Cid, Y. Underwater Pipe and Valve 3D Recognition Using Deep Learning Segmentation. *J. Mar. Sci. Eng.* **2021**, *9*, 5. https://dx.doi.org/10.3390/jmse9010005

Received: 10 December 2020
Accepted: 18 December 2020
Published: 23 December 2020

1. Introduction

During the past few decades, the interest in underwater intervention has grown exponentially as more often it is necessary to perform underwater tasks like surveying, sampling, archaeology exploration or industrial infrastructure inspection and maintenance of offshore oil and gas structures, submerged oil wells or pipeline networks, among others [1–5].

Historically, scuba diving has been the prevailing method of conducting the aforementioned tasks. However, performing these missions in a harsh environment like open water scenarios is slow, dangerous, and resource consuming. More recently, thanks to technological advances such as Remotely Operated Vehicles (*ROVs*) equipped with manipulators, more deep and complex underwater scenarios are accessible for scientific and industrial activities.

Nonetheless, these *ROVs* have complex dynamics that make their piloting a difficult and error-prone task, requiring trained operators. In addition, these vehicles require a support vessel, which leads to expensive operational costs. To mitigate that, some research centres have started working towards intervention Autonomous Underwater Vehicles (*AUVs*) [6–8]. In addition, due to the complexity of the Underwater Vehicle Manipulator Systems (UVMS), recent studies have been published towards its control [9,10].

Traditionally, when operating in unknown underwater environments, acoustic bathymetric maps are used to get a first identification of the environment. Once the bathymetric information is available, *ROVs* or *AUVs* can be sent to obtain more detailed information using short distance sensors with higher resolution. The two main perception modalities used at close range are laser and video, thanks to their high resolution. They are used during the approach, object recognition and intervention phases. Existing solutions for all perception modalities are reviewed in Section 2.1.

The underwater environment is one of the most problematic in terms of sensing in general and in terms of object perception in particular. The main challenges of underwater perception include distortion in signals, light propagation artefacts like absorption and scattering, water turbidity changes or depth-depending colour distortion.

Accurate and robust object detection, identification of target objects in different experimental conditions and pose estimation are essential requirements for the execution of manipulation tasks.

In this work, we propose a deep learning based approach to recognise pipes and valves in multiple underwater scenarios , using the 3D RGB point cloud information provided by a stereo camera, for real-time *AUV* inspection and manipulation tasks.

The remainder of this paper is structured as follows: Section 2 reviews related work on underwater perception and pipe and valve identification and highlights the main contributions of this work. Section 3 describes the adopted methodology and materials used in this study. The experimental results are presented and discussed in Section 4. Finally, Section 5 outlines the main conclusions and future work.

2. Related Work and Contributions

2.1. State of the Art

Even though computer vision is one of the most complete and used perception modalities in robotics and object recognition tasks, it has not been widely used in underwater scenarios. Light transmission problems and water turbidity affect the images clarity, colouring and produce distortions; these factors have favoured the usage of other perception techniques.

Sonar sensing has been largely used for object localisation or environment identification in underwater scenarios [11,12]. In [13], Kim et al. present an AdaBoost based method for underwater object detection, while Wang et al. [14] propose a combination of non-local spatial information and frog leaping algorithm to detect underwater objects in sonar images. More recently, object detection deep learning techniques have started to apply over sonar imaging in applications such as detection of underwater bodies in [15,16] or underwater mine detection in [17]. Sonar imaging also presents some drawbacks as it tends to generate noisy images, losing texture information; and are not capable of gathering colour information, which is useful in object recognition tasks.

Underwater laser scans are another perception technique used for object recognition, providing accurate 3D data. In [18], Palomer et al. present the calibration and integration of a laser scanner on an AUV for object manipulation. Himri et al. [19,20] use the same system to detect objects using a recognition and pose estimation pipeline based on point cloud matching. Inzartsev et al. [21] simulate the use of a single beam laser paired with a camera to capture its deformation and track an underwater pipeline. Laser scans are also affected by light transmission problems, have a very high initial cost and can only provide colourless point clouds.

The only perception modality that allows gathering of colour information for the scene is computer vision. Furthermore, some of its aforementioned weaknesses can be mitigated by adapting to the environmental conditions, adjusting the operation range, calibrating the cameras or colour correcting the obtained images.

Traditional computer vision approaches have been used to detect and track submerged artifacts [22–25], cables [26–28] and even pipelines [28–31]. Some works are based on shape and texture descriptors [28,31] or template matching [32,33], while others exploit colour segmentation to find regions of interest in the images, which are later further processed [25,34].

On pipeline detection, Kallasi et al. in [35] and Razzini et al. in [7,36] present traditional computer vision methods combining shape and colouring information to detect pipes in underwater scenarios and later project them into point clouds obtained from stereo vision. In these works, the point cloud information is not used to assist the pipe recognition process.

The first found trainable system to detect pipelines is presented in [37] by Rekik et al. using the objects structure and content features along a Support Vector Machine to classify between positive and negative underwater pipe images samples. Later, Nunes et. al introduced the application of a Convolutional Neural Network in [38] to classify up to five underwater objects, including a pipeline. In both of these works, no position of the object is given, but simply a binary output on the object's presence.

The application of computer vision approaches based on deep learning in underwater scenarios has been limited to the detection and pose estimation of 3D-printed objects in [39] or for living organisms detection like fishes [40] or jellyfishes [41]. Few research studies involving pipelines are restricted to damage evaluation [42,43] or valve detection for navigation [44] working with images taken from inside the pipelines. The only known work addressing pipeline recognition using deep learning is from Guerra et al. in [45], where a camera-equipped drone is used to detect pipelines in industrial environments.

To the best knowledge of the authors, there are not works applying deep learning techniques in underwater computer vision pipeline and valve recognition, nor implementing the usage of point cloud information on the detection process itself.

2.2. Main Contributions

The main contributions of this paper are composed of:

1. Generation of a novel point cloud dataset containing pipes and different types of valves in varied underwater scenarios, providing enough data to perform a robust training and testing of the selected deep neural network.
2. Implementation and testing of the PointNet architecture in underwater environments to detect pipes and valves.
3. Studying the suitability of the PointNet network on real-time autonomous underwater recognition tasks in terms of detection performance and inference time by tuning diverse hyperparameter values.
4. The datasets (point clouds and corresponding ground truths) along with a trained model are provided to the scientific community.

3. Materials and Methods

This section presents an overview of the selected network; explains the acquisition, labelling and organisation of the data; and details the studied network hyperparameters, the validation process and the evaluation metrics.

3.1. Deep Learning Network

To perform the pipe and valve 3D recognition from point cloud segmentation, we selected the PointNet deep neural network [46]. This is a unified architecture for applications ranging from object classification and part segmentation to scene semantic segmentation. PointNet is a highly efficient and effective network, obtaining great metrics in both object classification and segmentation tasks in indoor and outdoor scenarios [46]. However, it has never been tested in underwater scenarios. The whole PointNet architecture is shown in Figure 1.

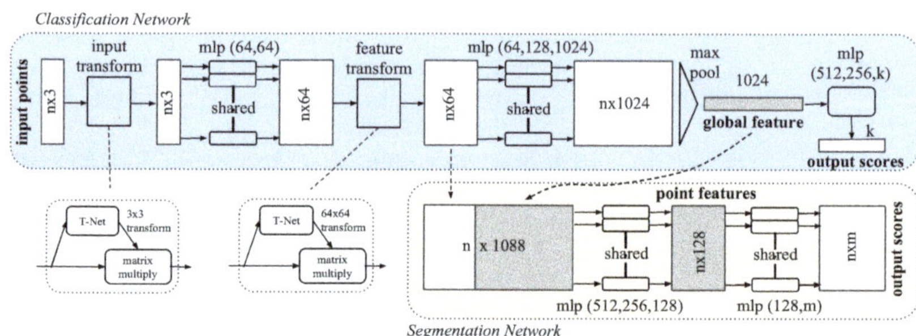

Figure 1. PointNet architecture. Reproduced from [46], with permission from publisher Hao Su, 2020.

In this paper, we use the *Segmentation Network* of PointNet. This network is an extension to the *Classification Network*, as it can be seen in Figure 1. Some of its key features include:

- The integration of max pooling layers as symmetric function to aggregate the information from each point, making the model invariant to input permutations.
- Being able to predict per point features that rely both on local structures from nearby points and global information which makes the prediction invariant to object transformations such as translations or rotations. This combination of local and global information is obtained by concatenating the global point cloud feature vector with the local per point features.
- Making the semantic labeling of a point cloud invariant to the point cloud geometric transformations by aligning all input set to a canonical space before feature extraction. To achieve this, an affine transformation matrix is predicted using a mini-network (T-net in Figure 1) and directly applied to the coordinates of input points.

The PointNet architecture takes as input point clouds and it outputs a class label for each point. During the training, the network is also fed with ground truth point clouds, where each point is labelled with its pertaining class. The labelling process is further detailed in Section 3.2.2.

As the original PointNet implementation, we used a softmax cross-entropy loss along an Adam optimiser. The decay rate for batch normalisation starts with 0.5 and is gradually increased to 0.99. In addition, we applied a dropout with keep ratio 0.7 on the last fully connected layer, before class score prediction. Other hyperparameters values such as learning rate or batch size are discussed, along other parameters, on Section 3.3.

Furthermore, to improve the network performance, we implemented an early stopping strategy based on the work of Prechelt in [47], assuring that the network training process stops at an epoch that ensures minimum divergence between validation and training losses. This technique allows for obtaining a more general and broad training, avoiding overfitting.

3.2. Data

This subsection explains the acquisition, labelling and organisation of the data used to train and test the PointNet neural network.

3.2.1. Acquisition

As mentioned in Section 3.1, the PointNet uses pointclouds for its training and inference. To obtain the point clouds, we set up a Bumblebee2 Firewire stereo rig [48] on an Autonomous Surface Vehicle (ASV) through a *Robot Operating System* (ROS) framework.

First, we calibrated the stereo rig both on fresh and salt water using the ROS package *image_pipeline/camera_calibration* [49,50]. It uses a chessboard pattern to obtain the camera, rectification and projection matrices along the distortion coefficients for both cameras.

The acquired synchronised pairs of left-right images (resolution: 1024 × 768 pixels) are processed as follows by the *image_pipeline/stere_image_proc* ROS package [51] to calculate the disparity between pairs of images based on epipolar matching [52], obtaining the corresponding depth of each pixel from the stereo rig.

Finally, combining this depth information with the RGB colouring from the original images, we generate the point clouds. An example of the acquisition is pictured in Figure 2.

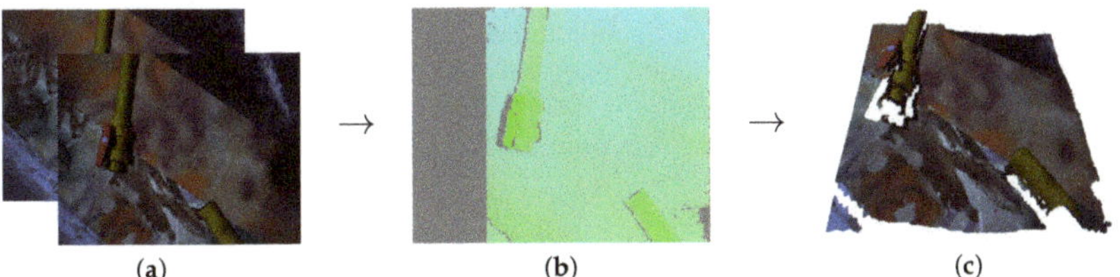

Figure 2. Data acquisition process. (**a**) left and right stereo images, (**b**) disparity image, (**c**) point cloud.

3.2.2. Ground Truth Labelling

Ground truth annotations are manually built from the point clouds, where the pixels corresponding to each class are marked with a different label. The studied classes and their RGB labels are: *Pipe* (0, 255, 0), *Valve* (0, 0, 255) and *Background* (0, 0, 0). Figure 3 shows a couple of point clouds along with their corresponding ground truth annotations.

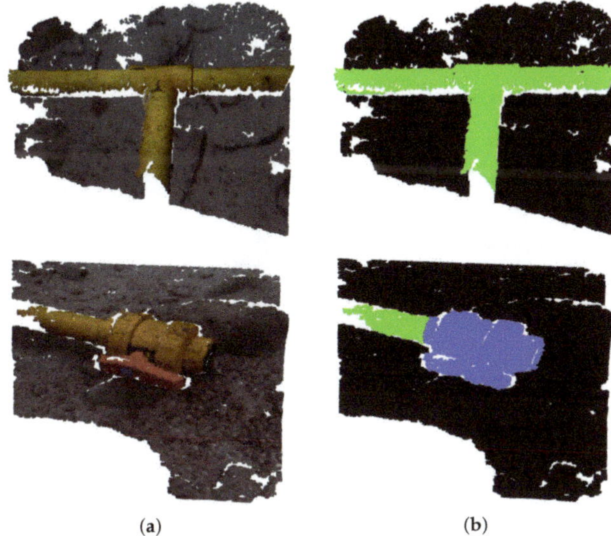

Figure 3. (**a**) Original point cloud; (**b**) ground truth annotations, points corresponding to pipes are marked in green; to valves, in blue; and to background, in black.

3.2.3. Dataset Managing

Following the steps described in the previous section, we generated two datasets. The first one includes a total of 262 point clouds along with their ground truths. It was obtained on an artificial pool and contains diverse connections between pipes of different diameters and 2/3 way valves. It also contains other objects such as cement blocks and ceramic vessels, always over a plastic sheeting simulating different textures. This dataset is split into a train-validation set (90% of the data, 236 point clouds) and a test set (10% of the data, 26 point clouds). The different combinations of elements and textures increase its diversity, helping to assure the robustness in the training and reduce overfitting. From now on, we will refer to this dataset as the *Pool* dataset.

The second dataset includes a total of 22 point clouds and their corresponding ground truths. It was obtained in the sea and contains different pipe connections and valves positions. In addition, these 22 point clouds were obtained over diverse types of seabed, such as sand, rocks, algae, or a combination of them. This dataset is used to perform a secondary test, as it contains point clouds with different characteristics of the ones used to train and validate the network, allowing us to assess how well the network generalises its training to new conditions. From now on, we will refer to this dataset as the *Sea* dataset.

Figure 4 illustrates the dataset managing, while in Figure 5 some examples of point clouds from both datasets are shown.

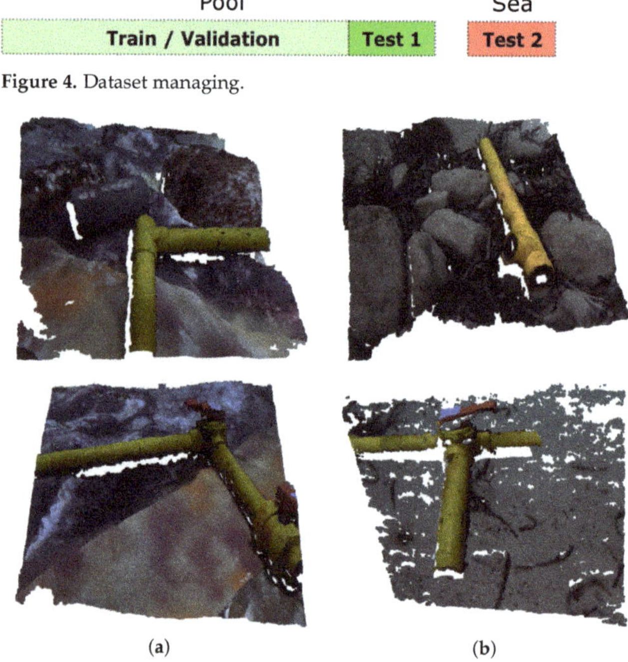

Figure 4. Dataset managing.

(**a**) (**b**)

Figure 5. Examples of point clouds from (**a**) *Pool* dataset and (**b**) *Sea* dataset.

3.3. Hyperparameter Study

When training a neural network, there are hyperparameters which can be tuned, changing some of the features of the network or the training process itself. We selected some of these hyperparameters and trained the network using different values to study their effect over its performance in underwater scenarios. The considered hyperparameters were:

- Batch size: number of training samples utilised in one iteration before backpropagating.
- Learning rate: affects the size of the matrix changes that the network takes when searching for an optimal solution.
- Block (B) and stride (S) size: to prepare the network input, the point clouds are sampled into blocks of BxB meters, with a sliding window of stride S meters.
- Number of points: maximum number of allowed points per block. If it exceeds, random points are deleted. Used to control the point cloud density.

The tested values for each hyperparameter are shown in Table 1. In total, 13 experiments are conducted, one using the hyperparameter values used in the original PointNet implementation [46] (marked in bold in Table 1); and 12 more, each one fixing three of the aforementioned hyperparameters to their original values and using one of the other tested values for the fourth hyperparameter. This way, the effect of each hyperparameter and its value over the performance is isolated.

Table 1. Tested hyperparameter values. Original values are marked in bold.

Hyperparameter	Tested Values					
Batch size	16	24	**32**			
Learning rate	0.005	**0.001**	0.0002			
Block-stride	2-2	2-1	**1-1**	1-0.75		
Num. points	**4096**	2048	1024	512	256	128

3.4. Validation

3.4.1. Validation Process

To ensure the robustness of the results generated for the 13 experiments, we used the 10 k-fold cross-validation method [53]. Using this method, the train-validation set of the *Pool* dataset is split into ten equally sized subsets. The network is trained ten times as follows, each one using a different subset as validation (23 point clouds) and the nine remaining as training (213 point clouds), generating ten models which are tested against both *Pool* and *Sea* test sets. Finally, each experiment performance is computed as the mean of the results of its 10 cross-validation models. This method reduces the variability of the results, as these are less dependent on the selected training and validation subsets, therefore obtaining a more accurate performance estimation. Figure 6 depicts the k-fold cross-validation technique applied to the dataset managing described in Section 3.2.3

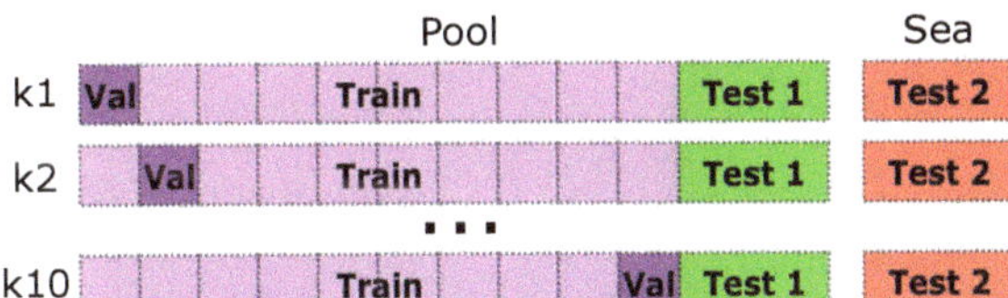

Figure 6. Implementation of the 10k-fold cross-validation method.

3.4.2. Evaluation Metrics

To evaluate a model performance, we make a point-wise comparison between its predictions and their corresponding ground truth annotations, generating a multi-class confusion matrix. This confusion matrix indicates, for each class: the number of points correctly identified belonging to that class, *True Positives* (TP) and not belonging to it, *True Negatives* (TN); the number of points misclassified as the studied class, *False Positives* (FP); and the number of points belonging to that class misclassified as another one, *False Negatives* (FN). Finally, the TP, FP and FN values are used to calculate the *Precision*, *Recall* and *F1-score* for each class, following Equations (1)–(3):

$$Precision = \frac{TP}{TP + FP} \, , \tag{1}$$

$$Recall = \frac{TP}{TP + FN} \, , \tag{2}$$

$$F1\text{-}score = 2 \times \frac{Recall \times Precision}{Recall + Precision} \, . \tag{3}$$

Additionally, the mean time that a model takes to perform the inference of a point cloud is calculated. This metric is very important, as it defines the frequency that information is provided to the system. In underwater applications, it would directly affect the agility and responsiveness of the AUV that this network could be integrated in, having an impact over the final operation time.

4. Experimental Results and Discussion

This section reports the performance obtained for each experiment over the *Pool* and *Sea* test sets and discusses the effect of each hyperparameter over it. The notation used to name each experiment corresponds as follows: "Base" for the experiment conducted using the original hyperparameter values, marked in bold in Table 1; the other experiments are notated as an abbreviation of the modified hyperparameter for that experiment ("Batch" for batch size, "Lr" for learning rate, "BS" for block-stride and "Np" for number of points) followed by the actual value of the hyperparameter for that experiment. For instance, experiment *Batch 24* uses all original hyperparameter values except for the batch size, which in this case is 24.

4.1. Pool Dataset Results

Table 2 shows the *F1-scores* obtained for the studied classes and its mean for all experiments when evaluated over the *Pool* test set. The mean inference time for each experiment is showcased in Figure 7 as follows.

Table 2. *Pool* test set *F1-scores*.

Experiment	F1_Pipe	F1_Valve	F1_Background	F1_Mean
Base	97.0%	93.1%	99.8%	96.6%
Batch 24	96.8%	92.7%	99.8%	96.4%
Batch 16	96.7%	92.3%	99.8%	96.2%
Lr 0005	96.4%	91.0%	99.7%	95.7%
Lr 00002	96.5%	92.5%	99.7%	96.2%
BS 2_2	96.0%	90.8%	99.7%	95.5%
BS 2_1	96.9%	93.3%	**99.8%**	96.7%
BS 1_075	**97.1%**	**94.9%**	99.7%	**97.2%**
Np 2048	96.7%	92.2%	99.8%	96.2%
Np 1024	96.9%	93.2%	99.8%	96.6%
Np 512	96.8%	92.6%	99.8%	96.4%
Np 256	96.9%	93.4%	99.8%	96.7%
Np 128	96.7%	92.8%	99.8%	96.4%

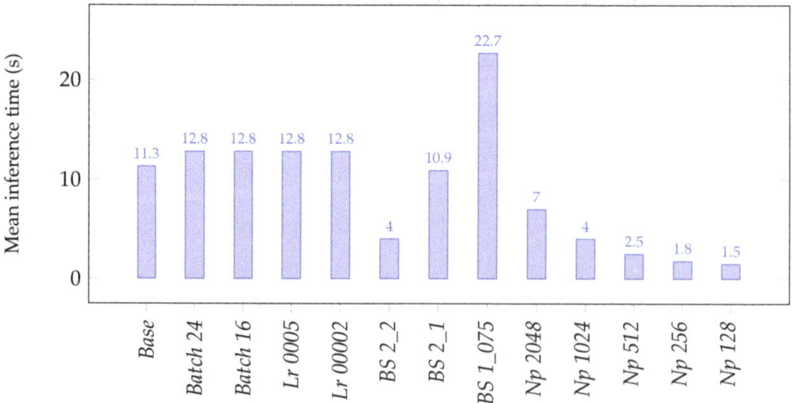

Figure 7. *Pool* test set mean inference time.

The results presented in Table 2 show that all experiments achieved a mean *F1-score* greater than 95.5%, with the highest value of 97.2% for the experiment *BS 1_075*, which has a smaller block stride than its size, overlapping information. Considering the figures of mean *F1-score* for all experiments, it is safe to say that no hyperparameter seemed to represent a major shift in the network behaviour.

Looking at the metrics presented by the best performing experiment for each class, it can be seen that the *Pipe* class achieved an *F1-score* of 97.1%, outperforming other state-of-the-art methods for underwater pipe segmentation: [35]—traditional computer vision algorithms over 2D underwater images achieving an *F1-score* of 94.1%, [7]—traditional computer vision algorithms over 2D underwater images achieving a mean *F1-score* over three datasets of 88.0% and [45]—deep leaning approach for 2D drone imagery achieving a pixel-wise accuracy of 73.1%. For the valve class, the *BS 1_075* experiment achieved a *F1-score* of 94.9%, being a more challenging class due to its complex geometry. As far as the authors know, no comparable work on underwater valve detection has been identified. Finally, for the more prevailing *Background* class, the best performing experiment achieved an *F1-score* of 99.7%.

The results on mean inference time for each experiment presented in Figure 7 shows that the batch size and learning rate hyperparameter values do not influence the inference time or have little impact, as their value is very similar to the one obtained in the *Base* experiment. On the contrary, the block and stride size highly affect the inference time, the bigger the information block or the stride between blocks, the faster the network can analyse a point cloud, and vice versa. Finally, the maximum number of allowed points per block also has a direct impact over the inference time, the lower it is, the faster the network can analyse a point cloud, as it becomes less dense. The time analysis was carried out in a computer with the following specs—processor: Intel i7-7700, RAM: 16 GB, GPU: NVIDIA GeForce GTX 1080.

Taking into account both metrics, *BS 1_075* presented the best *F1-score* and has the highest inference time. In this experiment, the network uses a small block size and stride, being able to analyse the data and extract its features better, at the cost of taking longer. The hyperparameter values of this experiment are a good fit for a system in which quick responsiveness to changes and high frequency of information are not a priority, allowing for maximising the recognition performance.

On the other hand, experiments such as *BS 2_2* or *Np 1024, 512, 256, 128* were able to maintain very high *F1-scores* while significantly reducing the inference time. The hyperparameter values tested in these experiments are a good fit for more agile systems that need a higher frequency of information and responsiveness to changes.

Figure 8 shows some examples of original point clouds from the *Pool* test set along with their corresponding ground truth annotations and network predictions.

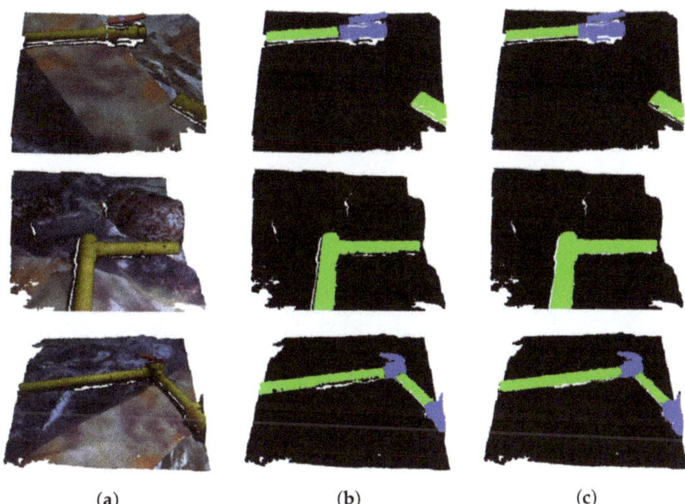

Figure 8. Qualitative results for the *Pool* test set. (**a**) original point cloud, (**b**) ground truth annotations, (**c**) network prediction.

4.2. Sea Dataset Results

Table 3 shows the *F1-scores* obtained for the studied classes and its mean, for all experiments when evaluated over the *Sea* test set. The mean inference time for each experiment is showcased in Figure 9 as follows.

Table 3. *Sea* test set *F1-scores*.

Experiment	F1_Pipe	F1_Valve	F1_Background	F1_Mean
Base	85.9%	79.5%	98.8%	88.1%
Batch 24	87.2%	79.9%	98.9%	88.7%
Batch 16	**88.1%**	80.9%	**99.0%**	**89.3%**
Lr 0005	86.2%	**81.2%**	98.8%	88.7%
Lr 00002	85.2%	76.3%	98.7%	86.8%
BS 2_2	80.7%	77.2%	**97.9%**	85.3%
BS 2_1	80.2%	79.7%	97.6%	85.8%
BS 1_075	86.7%	73.9%	99.0%	86.5%
Np 2048	85.2%	80.1%	98.5%	87.9%
Np 1024	86.1%	77.8%	98.8%	87.6%
Np 512	85.4%	70.7%	98.8%	85.0%
Np 256	87.1%	80.2%	98.9%	88.8%
Np 128	84.5%	71.5%	98.7%	84.9%

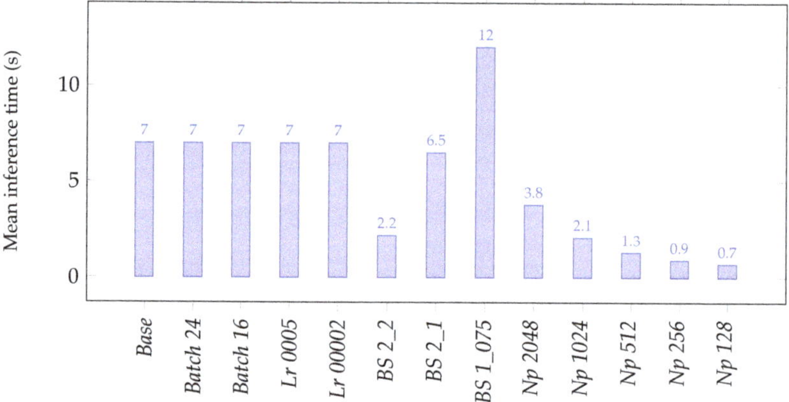

Figure 9. *Sea* test set mean inference time.

The results presented in Table 3 show that all experiments achieved a mean *F1-score* greater than 84.9% with the highest value of 89.3% for the experiment *Batch 16*. On average, the mean *F1-score* was around 9% lower than for the *Pool* test set. Even so, all experiments maintained high *F1-scores*. Again, the *F1-scores* of the *Pipe* and *Valve* classes are relatively lower than for the *Background* class. Even though the *Sea* test set is more challenging, as it contains unseen pipe and valve connections and environment conditions, the network was able to generalise its training and avoid overfitting.

The results on mean inference time for each experiment presented in Figure 9 shows that the mean inference times for the *Sea* test set are proportionally lower than the *Pool* test set for all experiments. This occurs because the *Sea* test set contains smaller point clouds with fewer points.

Figure 10 shows some examples of original point clouds from the *Sea* test set along with their corresponding ground truth annotations and network predictions.

Figure 10. Qualitative results for the *Sea* test set. (**a**) original point cloud; (**b**) ground truth annotations; (**c**) network prediction.

5. Conclusions and Future Work

This work studied the implementation of the PointNet deep neural network in underwater scenarios to recognise pipes and valves from point clouds. First, two datasets of point clouds were gathered, providing enough data for the training and testing of the network. From these, a train-validation set and two test sets were generated, a primary test set with similar characteristics as the training data and a secondary one containing unseen pipe and valve links and environment conditions to test the network training generalisation and overfitting. Then, diverse hyperparameter values were tested to study their effect over the network performance, both in the recognition task and inference time.

Results from the recognition task concluded that the network was able to identify pipes and valves with high accuracy for all experiments in both *Pool* and *Sea* test sets, reaching *F1-scores* of 97.2% and 89.3%, respectively. Regarding the network inference time, results showed that it is highly dependent on the size of information block and its stride; and to the point clouds density.

From the performed experiments, we obtained a range of models covering different trade-offs between detection performance and inference time, enabling the network implementation into a wider spectrum of systems, adapting to its detection and computational cost requirements. The *BS 1_075* experiment presented metrics that fitted a slower, more still system, while experiments like *BS 2_2* or *Np 1024, 512, 256, 128* are a good fit for more agile and dynamic systems.

The implementation of the PointNet network in underwater scenarios presented some challenges, like ensuring its recognition performance when trained with point clouds obtained from underwater images, and its suitability to be integrated on an AUV due to its computational cost. With the results obtained in this work, we have demonstrated the validity of the PointNet deep neural network to detect pipes and valves in underwater scenarios for AUV manipulation and inspection tasks.

The datasets and code, along with one of the *Base* experiment trained models, are publicly available at http://srv.uib.es/3d-pipes-1/ (UIB-SRV-3D-pipes) for the scientific community to test or replicate our experiments.

Further steps need to be taken in order to achieve an underwater object localisation and positioning for ROV and AUV intervention using the object recognition presented in this work. We propose the following future work:

1. Performing an instance-based detection from the presented pixel-based one, allowing for recognition of pipes and valves as a whole object and to classify them by type (two or three way) or status (opened or closed).
2. Using the depth information provided by the stereo cameras along with the instance detection to achieve a spatial 3D positioning of each object. Once the network is implemented in an AUV, this would provide the vehicle with the information to manipulate and intervene with the recognised objects.

Author Contributions: Conceptualisation, G.O.-C. and Y.G.-C.; methodology, M.M.-A.; software, M.M.-A., M.P.-M. and A.M.-T.; validation, M.M.-A.; investigation, M.M.-A. and M.P.-M.; resources, G.O.-C. and Y.G.-C.; data curation, M.M.-A., M.P.-M. and A.M.-T.; writing—original draft preparation, M.M.-A. and M.P.-M.; writing—review and editing, M.M.-A., M.P.-M., A.M.-T., G.O.-C. and Y.G.-C.; supervision, Y.G.-C.; project administration, G.O.-C. and Y.G.-C.; funding acquisition, G.O.-C. and Y.G.-C. All authors have read and agreed to the published version of the manuscript.

Funding: Miguel Martin-Abadal was supported by the Ministry of Economy and Competitiveness (AEI,FEDER,UE), under contract DPI2017-86372-C3-3-R. Gabriel Oliver-Codina was supported by Ministry of Economy and Competitiveness (AEI,FEDER,UE), under contract DPI2017-86372-C3-3-R. Yolanda Gonzalez-Cid was supported by the Ministry of Economy and Competitiveness (AEI,FEDER,UE), under contracts TIN2017-85572-P and DPI2017-86372-C3-3-R; and by the Comunitat Autonoma de les Illes Balears through the Direcció General de Política Universitaria i Recerca with funds from the Tourist Stay Tax Law (PRD2018/34).

Data Availability Statement: Publicly available datasets were analysed in this study. This data can be found here: http://srv.uib.es/3d-pipes-1/.

Conflicts of Interest: The authors declare no conflict of interest.

References

1. Yu, M.; Ariamuthu Venkidasalapathy, J.; Shen, Y.; Quddus, N.; Mannan, M.S. Bow-tie Analysis of Underwater Robots in Offshore Oil and Gas Operations. In Proceedings of the Offshore Technology Conference, Houston, TX, USA, 1–4 May 2017. [CrossRef]
2. Costa, M.; Pinto, J.; Ribeiro, M.; Lima, K.; Monteiro, A.; Kowalczyk, P.; Sousa, J. Underwater Archaeology with Light AUVs. In Proceedings of the OCEANS 2019—Marseille, Marseille, France, 17–20 June 2019; pp. 1–6. doi:10.1109/OCEANSE.2019.8867503. [CrossRef]
3. Asakawa, K.; Kojima, J.; Kato, Y.; Matsumoto, S.; Kato, N. Autonomous underwater vehicle AQUA EXPLORER 2 for inspection of underwater cables. In Proceedings of the 2000 International Symposium on Underwater Technology (Cat. No.00EX418), Tokyo, Japan, 26 May 2000; pp. 242–247. [CrossRef]
4. Jacobi, M.; Karimanzira, D. Underwater pipeline and cable inspection using autonomous underwater vehicles. In Proceedings of the 2013 MTS/IEEE OCEANS—Bergen, Bergen, Norway, 10–14 June 2013; pp. 1–6. [CrossRef]
5. Capocci, R.; Dooly, G.; Omerdić, E.; Coleman, J.; Newe, T.; Toal, D. Inspection-Class Remotely Operated Vehicles—A Review. *J. Mar. Sci. Eng.* **2017**, *5*, 13. [CrossRef]
6. Ridao, P.; Carreras, M.; Ribas, D.; Sanz, P.J.; Oliver, G. Intervention AUVs: The Next, Challenge. *Annu. Rev. Control* **2015**, *40*, 227–241. [CrossRef]
7. Lodi Rizzini, D.; Kallasi, F.; Aleotti, J.; Oleari, F.; Caselli, S. Integration of a stereo vision system into an autonomous underwater vehicle for pipe manipulation tasks. *Comput. Electr. Eng.* **2017**, *58*, 560–571. [CrossRef]
8. Heshmati-Alamdari, S.; Nikou, A.; Dimarogonas, D.V. Robust Trajectory Tracking Control for Underactuated Autonomous Underwater Vehicles in Uncertain Environments. *IEEE Trans. Autom. Sci. Eng.* **2020**, 1–14. [CrossRef]
9. Nikou, A.; Verginis, C.K.; Dimarogonas, D.V. A Tube-based MPC Scheme for Interaction Control of Underwater Vehicle Manipulator Systems. In Proceedings of the 2018 IEEE/OES Autonomous Underwater Vehicle Workshop (AUV), Porto, Portugal, 6–9 November 2018; pp. 1–6. [CrossRef]
10. Heshmati-Alamdari, S.; Bechlioulis, C.P.; Karras, G.C.; Nikou, A.; Dimarogonas, D.V.; Kyriakopoulos, K.J. A robust interaction control approach for underwater vehicle manipulator systems. *Annu. Rev. Control* **2018**, *46*, 315–325. [CrossRef]
11. Jonsson, P.; Sillitoe, I.; Dushaw, B.; Nystuen, J.; Heltne, J. Observing using sound and light—A short review of underwater acoustic and video-based methods. *Ocean Sci. Discuss.* **2009**, *6*, 819–870. [CrossRef]
12. Burguera, A.; Bonin-Font, F. On-Line Multi-Class Segmentation of Side-Scan Sonar Imagery Using an Autonomous Underwater Vehicle. *J. Mar. Sci. Eng.* **2020**, *8*, 557. [CrossRef]

13. Kim, B.; Yu, S. Imaging sonar based real-time underwater object detection utilizing AdaBoost method. In Proceedings of the 2017 IEEE Underwater Technology (UT), Busan, Korea, 21–24 Februry 2017; Volume 845, pp. 1–5. [CrossRef]
14. Wang, X.; Liu, S.; Liu, Z. Underwater sonar image detection: A combination of nonlocal spatial information and quantum-inspired shuoed frog leaping algorithm. *PLoS ONE* **2017**, *12*, e0177666. [CrossRef]
15. Lee, S.; Park, B.; Kim, A. Deep Learning from Shallow Dives: Sonar Image Generation and Training for Underwater Object Detection. *arXiv* **2018**, arXiv:1810.07990.
16. Lee, S.; Park, B.; Kim, A. A Deep Learning based Submerged Body Classification Using Underwater Imaging Sonar. In Proceedings of the 2019 16th International Conference on Ubiquitous Robots (UR), Jeju, Korea, 24–27 June 2019; pp. 106–112. [CrossRef]
17. Denos, K.; Ravaut, M.; Fagette, A.; Lim, H. Deep learning applied to underwater mine warfare. In Proceedings of the OCEANS 2017—Aberdeen, Aberdeen, UK, 19–22 June 2017; pp. 1–7. [CrossRef]
18. Palomer, A.; Ridao, P.; Youakim, D.; Ribas, D.; Forest, J.; Petillot, Y. 3D laser scanner for underwater manipulation. *Sensors* **2018**, *18*, 1086. [CrossRef]
19. Himri, K.; Pi, R.; Ridao, P.; Gracias, N.; Palomer, A.; Palomeras, N. Object Recognition and Pose Estimation using Laser scans for Advanced Underwater Manipulation. In Proceedings of the 2018 IEEE/OES Autonomous Underwater Vehicle Workshop (AUV), Porto, Portugal, 6–9 November 2018; pp. 1–6. [CrossRef]
20. Himri, K.; Ridao, P.; Gracias, N. 3D Object Recognition Based on Point Clouds in Underwater Environment with Global Descriptors: A Survey. *Sensors* **2019**, *19*, 4451. [CrossRef] [PubMed]
21. Inzartsev, A.; Eliseenko, G.; Panin, M.; Pavin, A.; Bobkov, V.; Morozov, M. Underwater pipeline inspection method for AUV based on laser line recognition: Simulation results. In Proceedings of the 2019 IEEE International Underwater Technology Symposium, UT 2019—Proceedings, Kaohsiung, Taiwan, 16–19 April 2019; pp. 1–8. [CrossRef]
22. Olmos, A.; Trucco, E. Detecting man-made objects in unconstrained subsea videos. In Proceedings of the British Machine Vision Conference, Cardiff, UK, 2–5 September 2002; pp. 50.1–50.10. [CrossRef]
23. Chen, Z.; Wang, H.; Xu, L.; Shen, J. Visual-adaptation-mechanism based underwater object extraction. *Opt. Laser Technol.* **2014**, *56*, 119–130. [CrossRef]
24. Ahmed, S.; Khan, M.F.R.; Labib, M.F.A.; Chowdhury, A.E. An Observation of Vision Based Underwater Object Detection and Tracking. In Proceedings of the 2020 3rd International Conference on Emerging Technologies in Computer Engineering: Machine Learning and Internet of Things (ICETCE), Jaipur, India, 7–8 February 2020; pp. 117–122. [CrossRef]
25. Prats, M.; García, J.C.; Wirth, S.; Ribas, D.; Sanz, P.J.; Ridao, P.; Gracias, N.; Oliver, G. Multipurpose autonomous underwater intervention: A systems integration perspective. In Proceedings of the 2012 20th Mediterranean Conference on Control Automation (MED), Barcelona, Spain, 3–6 July 2012; pp. 1379–1384. [CrossRef]
26. Ortiz, A.; Simó, M.; Oliver, G. A vision system for an underwater cable tracker. *Mach. Vis. Appl.* **2002**, *13*, 129–140. [CrossRef]
27. Fatan, M.; Daliri, M.R.; Mohammad Shahri, A. Underwater cable detection in the images using edge classification based on texture information. *Meas. J. Int. Meas. Confed.* **2016**, *91*, 309–317. [CrossRef]
28. Narimani, M.; Nazem, S.; Loueipour, M. Robotics vision-based system for an underwater pipeline and cable tracker. In Proceedings of the OCEANS 2009-EUROPE, Bremen, Germany, 11–14 May 2009; pp. 1–6. [CrossRef]
29. Tascini, G.; Zingaretti, P.; Conte, G. Real-time inspection by submarine images. *J. Electron. Imaging* **1996**, *5*, 432–442. [CrossRef]
30. Zingaretti, P.; Zanoli, S.M. Robust real-time detection of an underwater pipeline. *Eng. Appl. Artif. Intell.* **1998**, *11*, 257–268. [CrossRef]
31. Foresti, G.L.; Gentili, S. A hierarchical classification system for object recognition in underwater environments. *IEEE J. Ocean. Eng.* **2002**, *27*, 66–78. [CrossRef]
32. Kim, D.; Lee, D.; Myung, H.; Choi, H. Object detection and tracking for autonomous underwater robots using weighted template matching. In Proceedings of the 2012 Oceans—Yeosu, Yeosu, Korea, 21–24 May 2012; pp. 1–5. [CrossRef]
33. Lee, D.; Kim, G.; Kim, D.; Myung, H.; Choi, H.T. Vision-based object detection and tracking for autonomous navigation of underwater robots. *Ocean Eng.* **2012**, *48*, 59–68. [CrossRef]
34. Bazeille, S.; Quidu, I.; Jaulin, L. Color-based underwater object recognition using water light attenuation. *Intell. Serv. Robot.* **2012**, *5*, 109–118. [CrossRef]
35. Kallasi, F.; Oleari, F.; Bottioni, M.; Lodi Rizzini, D.; Caselli, S. Object Detection and Pose Estimation Algorithms for Underwater Manipulation. In Proceedings of the 2014 Conference on Advances in Marine Robotics Applications, Palermo, Italy, 16–19 June 2014.
36. Lodi Rizzini, D.; Kallasi, F.; Oleari, F.; Caselli, S. Investigation of Vision-based Underwater Object Detection with Multiple Datasets. *Int. J. Adv. Robot. Syst.* **2015**, *12*, 1–13. [CrossRef]
37. Rekik, F.; Ayedi, W.; Jallouli, M. A Trainable System for Underwater Pipe Detection. *Pattern Recognit. Image Anal.* **2018**, *28*, 525–536. [CrossRef]
38. Nunes, A.; Gaspar, A.R.; Matos, A. Critical object recognition in underwater environment. In Proceedings of the OCEANS 2019—Marseille, Marseille, France, 17–20 June 2019; pp. 1–6. [CrossRef]
39. Jeon, M.; Lee, Y.; Shin, Y.S.; Jang, H.; Kim, A. Underwater Object Detection and Pose Estimation using Deep Learning. *IFAC-PapersOnLine* **2019**, *52*, 78–81. [CrossRef]
40. Jalal, A.; Salman, A.; Mian, A.; Shortis, M.; Shafait, F. Fish detection and species classification in underwater environments using deep learning with temporal information. *Ecol. Informatics* **2020**, *57*, 101088. [CrossRef]

41. Martin-Abadal, M.; Ruiz-Frau, A.; Hinz, H.; Gonzalez-Cid, Y. Jellytoring: Real-time jellyfish monitoring based on deep learning object detection. *Sensors* **2020**, *20*, 1708. [CrossRef] [PubMed]
42. Kumar, S.S.; Abraham, D.M.; Jahanshahi, M.R.; Iseley, T.; Starr, J. Automated defect classification in sewer closed circuit television inspections using deep convolutional neural networks. *Autom. Constr.* **2018**, *91*, 273–283. [CrossRef]
43. Cheng, J.C.; Wang, M. Automated detection of sewer pipe defects in closed-circuit television images using deep learning techniques. *Autom. Constr.* **2018**, *95*, 155–171. [CrossRef]
44. Rayhana, R.; Jiao, Y.; Liu, Z.; Wu, A.; Kong, X. Water pipe valve detection by using deep neural networks. In *Smart Structures and NDE for Industry 4.0, Smart Cities, and Energy Systems*; SPIE: Bellingham, WA, USA, 2020; Volume 11382, pp. 20–27. [CrossRef]
45. Guerra, E.; Palacin, J.; Wang, Z.; Grau, A. Deep Learning-Based Detection of Pipes in Industrial Environments. In *Industrial Robotics*; IntechOpen: London, UK, 2020; doi:10.5772/intechopen.93164. [CrossRef]
46. Charles, R.Q.; Su, H.; Kaichun, M.; Guibas, L.J. PointNet: Deep Learning on Point Sets for 3D Classification and Segmentation. In Proceedings of the 2017 IEEE Conference on Computer Vision and Pattern Recognition (CVPR), Honolulu, HI, USA, 21–26 July 2017; pp. 77–85. [CrossRef]
47. Prechelt, L. Early Stopping—However, When? In *Neural Networks: Tricks of the Trade*, 2nd ed.; Springer: Berlin/Heidelberg, Germany, 2012; pp. 53–67._5. [CrossRef]
48. Bumblebee 2 Stereo Rig. Available online: https://www.flir.com/support/products/bumblebee2-firewire/#Overview (accessed on 7 December 2020).
49. ROS—Camera Calibration. Available online: http://wiki.ros.org/camera_calibration (accessed on 7 December 2020).
50. ROS—Camera Info. Available online: http://wiki.ros.org/image_pipeline/CameraInfo (accessed on 7 December 2020).
51. ROS—Stereo Image Proc. Available online: http://wiki.ros.org/stereo_image_proc (accessed on 7 December 2020).
52. Hartley, R.; Zisserman, A. *Multiple View Geometry in Computer Vision*, 2nd ed.; Cambridge University Press: Cambridge, MA, USA, 2003.
53. Geisser, S. The predictive sample reuse method with applications. *J. Am. Stat. Assoc.* **1975**, *70*, 320–328. [CrossRef]

Journal of
Marine Science and Engineering

Article

On-Line Multi-Class Segmentation of Side-Scan Sonar Imagery Using an Autonomous Underwater Vehicle

Antoni Burguera *,† and Francisco Bonin-Font †

Departament de Matemàtiques i Informàtica, Universitat de les Illes Balears, Carretera de Valldemossa Km. 7.5, 07122 Palma, Spain
* Correspondence: antoni.burguera@uib.es
† These authors contributed equally to this work.

Received: 8 July 2020; Accepted: 22 July 2020; Published: 24 July 2020

Abstract: This paper proposes a method to perform on-line multi-class segmentation of Side-Scan Sonar acoustic images, thus being able to build a semantic map of the sea bottom usable to search loop candidates in a SLAM context. The proposal follows three main steps. First, the sonar data is pre-processed by means of acoustics based models. Second, the data is segmented thanks to a lightweight Convolutional Neural Network which is fed with acoustic swaths gathered within a temporal window. Third, the segmented swaths are fused into a consistent segmented image. The experiments, performed with real data gathered in coastal areas of Mallorca (Spain), explore all the possible configurations and show the validity of our proposal both in terms of segmentation quality, with per-class precisions and recalls surpassing the 90%, and in terms of computational speed, requiring less than a 7% of CPU time on a standard laptop computer. The fully documented source code, and some trained models and datasets are provided as part of this study.

Keywords: sonar; underwater robotics; acoustic image segmentation; neural network

---

1. Introduction

Even though cameras are gaining popularity in underwater robotics, computer vision still presents some problems in these scenarios [1]. The particularities of the aquatic medium, such as light absorption, back scatter or flickering, among many others, significantly reduce the visibility range and the quality of the image. Because of that, underwater vision is usually constrained to missions in which the *Autonomous Underwater Vehicle* (AUV) can navigate close to the sea bottom to properly observe it [2,3].

To the contrary, acoustic sensors or *sonars* [4] are particularly well suited for subsea environments not only because of their large sensing range, but also because they are not influenced by the illumination conditions and thus they can operate easily in a wider range of scenarios. Whereas underwater cameras can observe objects that are a few meters away, sonars reach much larger distances. For example, the *Geological LOng-Range Inclined ASDIC* (GLORIA) operation range exceeds the 20 km [5,6]. That is why sonar is still the modality of choice in underwater robotics, being used as the main exteroceptive sensor [7] or combined with cameras for close range navigation.

There is a large variety of sonars ready to be used by an AUV. For example, the *Synthetic Aperture Sonar* (SAS) [8] is known to provide high resolution echo intensity profiles by gathering several measurements of each spot and fusing them during post-processing. Thanks to that, SAS are able to scan the sea bottom with resolutions far better than other sonars, reaching improvements of one or two orders of magnitude in the along-track direction. This advantage has a cost. One the one hand, using SAS constrains the maximum speed at which the AUV can move, since the same spot has to be observed several times. On the other hand, the mentioned high resolution depends on the AUV moving in straight trajectories, since observing the same spot from different angles may jeopardize the post-processing. Moreover, SAS are particularly expensive and their deployment is more complex than other types of sonar.

Another example is the *Mechanically Scanned Imaging Sonar* (MSIS) [9], whose most distinctive feature is its rotating sensing head which provides 360° echo intensity profiles of the environment. Because of that, this sensor is used to detect and map obstacles in the plane where the AUV navigates, though a few studies exist showing their use to scan the sea bottom [10]. The main drawback of this sensor is, precisely, the mechanical rotation which is responsible for very large scan times, usually between 10 and 20 seconds and also leads to a high power consumption. This also constrains the speed at which the AUV can move since moving at high speed would lead to distorted scans. Additionally, installing an MSIS on an AUV is not simple as they have a preferential mounting orientation.

The *Multi-Beam Sonars* (MBS) [11] sample a region of the sea bottom by emitting ultrasonic waves in an fan shape. The distance to the closest obstacles within their field of view is obtained by means of *Time of Flight* (TOF) techniques, thus computing the water depth. In contrast to other sonars, directional information from the returning sound waves is extracted using *beamforming* [12], so that a swath of depth readings is obtained from each single ping. This behaviour constitutes the MBS main advantage, as well as their most distinctive feature: contrarily to the previously mentioned sonars, MBS provide true 3D information of the ocean floor and, thus, they are commonly used to obtain subsea bathymetry. That is why they have been successfully applied to underwater mapping [13] and *Simultaneous Localization and Mapping* (SLAM) [14]. Their main disadvantages are their price, as well as, usually, their size and weight.

The *Side-Scan Sonar* (SSS) [15,16] provides echo intensity profiles similar to those of SAS and MSIS. The spatial resolution of SSS [17] is usually below that of SAS and, since they are not mounted on a rotating platform, they do not provide 360° views of the environment but slices of the sea floor. Moreover, they do not provide true bathymetry like MBS. In spite of these limitations when compared to SAS or MSIS, SSS are still the sensor of choice to obtain sea floor imagery, and they will probably remain in the near future for two main reasons.

On the one hand, SSS are economic, thus being suitable even in low cost robotics. On the other hand, they are particularly easy to deploy. They do not require any special mounting such as MBS or MSIS and they are even available as a towfish so they can be used without any additional infrastructure in some AUV and *Remotely Operated Vehicles* (ROV), as well as in ships. Also, their power consumption is below that of SAS, MSIS and MBS, thus being well suited in underwater robotics where the power tends to be a problem.

The most common application of SSS is to produce acoustic images of the sea bottom which are analysed off-line by humans. These images make it possible to detect some geological features [18] or to explore and analyse archaeological sites [19], among others, but mainly involving human analysis of the SSS data. Unfortunately, SSS imagery has not been traditionally used to perform autonomous navigation since the obtained acoustic images have some particularities that jeopardize their automatic analysis.

For example, since SSS measurements are slices of the sea bottom usually perpendicular to the motion direction, they do not overlap between them and, thus, they provide no information to directly estimate the AUV motion. Also, similarly to other sonars, SSS unevenly ensonify the targets, thus leading to echoes that do not only depend on the structure of the sea bottom but also on the particular ensonification pattern [17]. Moreover, since the hydrophone and the ultrasonic emitter are very close, the acoustic shadows, which correspond to occluded areas, strongly depend on the AUV position with respect to the target. This means that the same target leads to very different acoustic images depending on its position relative to the AUV. Finally, raw SSS images are geometrically distorted representations of the sea bottom [20] and properly correcting this distortion is a complex task [21].

There are only few studies dealing with these problems and pursuing fully automated SSS imagery analysis. Most of them either are too computationally demanding to be used on-line [22] or focus on areas with clearly distinguishable targets [23], thus lacking generality. Performing SLAM using SSS data is an almost unexplored research field and, at the extent of the authors knowledge, there are no studies fully performing SLAM with this type of sonar. For example, [24] proposes a target detection method from SSS data, but focuses on very specific, man made, environments. Also, [25] performs SLAM with SSS data in generic environments but still relies on hand labelled landmarks.

Automatic, on-line, analysis of SSS imagery is crucial to perform SLAM, which is necessary to build truly autonomous vehicles. SLAM relies on on-line place recognition, which consists on deciding whether the currently observed region was already observed in the past and constitute a so called *loop* or not. This process, usually referred to as *data registration*, can be extremely time consuming and error prone. Because of that, it is usual to pre-select some candidate loops with some fast algorithm and then perform data registration only with those candidates. This candidate selection could strongly benefit from an on-line segmentation of SSS data.

Accordingly, the first step towards robust place recognition for a fully operational SLAM approach using SSS can be to properly segment acoustic images into different classes. In this way, candidate loops could be searched at regions with overlapping classes and they could be subsequently refined to detect actual loops.

Properly segmenting SSS images on-line could be used in many other applications aside of SLAM, such as geological or biological submarine studies or archaeological research among many others. For example, an AUV in charge of measuring the coverage of a certain algae could be guided towards the boundaries of the regions classified as algae using the on-line segmented data.

Research on acoustic image segmentation is scarce and, similarly to previously mentioned studies, usually targeted at very particular and constrained scenarios [26,27] requiring high-resolution acoustic data [28,29]. Most of these studies rely on hand-crafted descriptors, often being constrained to a specific kind of environment.

For example, [29] specifically searches for shadows and edges and performs texture segmentation by means of the texture energy, thus relying on pre-defined hand-crafted descriptors of the environment that may only be suitable for a reduced range of environments. A similar situation can be found in [30], where an ad-hoc morphological filter to detect shadows is combined with erosions and dilations, or in [28], where the concept of *lacunarity* is used to segment SAS and SSS images. In all these cases good segmentation results are achieved but the texture segmentation methods, either hand-crafted or borrowed from the computer vision community, lack generality and constrain the applicability to certain types of scenarios. Moreover, most of these methods require large images to properly operate, thus jeopardizing their on-line application.

General purpose acoustic image segmentation is still an open field particularly challenging when it comes to SSS because of the above mentioned problems. Among these problems, the one of shadows leading to radically different images depending on the viewpoint is arguably the most difficult. Having objects of the same class, even the same object, with completely different features strongly increases the difficulties of any segmentation process. That is why several studies, such as the previously mentioned [29] or [30] as well as other studies targeting acoustic image matching [31] intentionally focus on detecting and dealing with the shadows.

Recent trends on image segmentation make use of *Neural Networks* (NN) [32]. In particular, *Convolutional Neural Networks* (CNN) have shown to provide exceptional results in front of situations which were extremely difficult to solve for traditional approaches, also providing a general solution to the segmentation problem. Unfortunately, NN in general and CNN in particular are said to have one important problem: they require large quantities of data to be trained. In most cases, such a large quantity of data is already available [33] and in some others the problem can be avoided taking advantage of transfer learning [34].

Unfortunately, when dealing with SSS, neither large quantities of data are available nor pre-trained NN can be used since they are commonly trained with terrestrial optical images and not with acoustic underwater data. For example, [35] proposes a NN approach to segment SSS data and has to pay special attention to data augmentation techniques in order to alleviate the lack of large training datasets.

Moreover, although NN are not necessarily slow after training, their computational requirements, both in terms of space and speed, may be a problem when it comes to AUVs where limitations in space and power supply prevent the use of fast computers endowed with *Graphics Processing Units* (GPU) or *Tensor Processing Units* (TPU). As a matter of fact, the previously mentioned study [35], which uses a well known NN architecture, even though it to pre-processes the data to reduce the NN computational requirements cannot be executed on-line.

The proposal in this paper is to overcome these problems by defining a CNN to segment SSS imagery not requiring large amounts of data to be trained and being fast enough to be deployed on-line on an AUV. To accomplish these goals, and being the main contributions of this paper, we:

- Derive an acoustics based method [17] to pre-process the data so that the NN has to deal with less uncertainties, thus facilitating its training and on-line usage.
- Propose a sliding window approach that makes it possible, when combined with the pre-processing, to train the NN with a small amount of data and to use it on-line even on AUVs with reduced computational power.
- Propose a Convolutional Neural Network following an encoder-decoder architecture in charge of segmenting the acoustic data.

Aside of these novelties, we present an additional contribution by releasing the fully documented source code, as well as different pre-trained models and some of the datasets used in the paper. All this code and data is available at https://github.com/aburguera/NNSSS.

This paper is structured as follows. First, the basics of SSS sensing are presented in Section 2. Afterwards, Section 3 describes how the SSS data is pre-processed based on underwater acoustics. Section 4 focuses on the proposed CNN. Both training and on-line usage are described as well as the proposed sliding window approach. Section 5 shows the experimental results, both those aimed at tuning the system and those devoted at evaluating its quality both quantitatively and qualitatively. Finally, Section 6 shows the main conclusions and provides an insight for further work.

2. The Side-Scan Sonar

2.1. Overview

A SSS is composed of two sensing heads, which are symmetrically mounted on the AUV on port and starboard. These sensing heads point at opposite directions perpendicular to the AUV motion direction while they observe the sea floor at a specific angle θ. This angle, which is called the *mounting angle*, is shown in Figure 1 together with the nomenclature, the whole setup and the symbols used throughout the paper.

Since the operation of the two sensing heads is identical, let us focus on one of them. One sensing head emits an ultrasonic pulse called *ping* at regular time intervals. This ultrasonic pulse not only moves along the sensor acoustic axis but also expands perpendicularly to it. This expansion is usually modelled by two angles called *openings*. The horizontal opening φ models the sound expansion in the horizontal plane XY as it moves over the Y axis. The vertical opening α models the sound expansion in the vertical plane YZ while it moves over the acoustic axis defined by θ.

The ultrasonic pulse will eventually collide with a region of the sea floor called the *ensonified region* (ER), which will partially scatter the pulse back to the sensor, where it will be analysed. The size of the ER depends, thus, on the openings and the altitude h at which the vehicle navigates. Typical SSS configurations involve large vertical openings α, of tens of degrees, and small horizontal openings φ of only a few degrees. This means that it is usually assumed that a ER is a thin strip of the ocean floor perpendicular to the AUV motion direction.

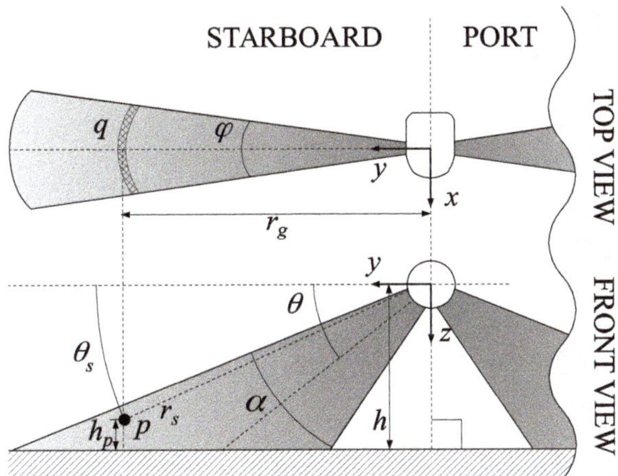

Figure 1. Side-scan sonar model. The x axis points to the AUV motion direction or along-track direction.

2.2. Sensor Operation

After emitting each ultrasonic pulse, the sensing head records the received echo intensities at fixed time intervals into a vector until a new pulse is emitted and the process starts again. Let this recorded data vector be referred to as a *swath*. Thus, a swath vector is obtained for each sonar ping. Each component of the swath, which holds information about the received echo intensity at a particular time step, is called a *bin*. That is, each swath is composed of several bins that can be seen as pixels of a one dimensional acoustic image.

Since the *slant range* r_s (see Figure 1) can be computed from the TOF of each bin, let us assume that each bin is associated to a particular distance from the sensor to the ocean floor. Changes in the speed of sound due to variations in water density, salinity or temperature, among others, are not taken into account in this paper. Accordingly, the speed at which bins are sampled is directly responsible for the slant range resolution δ_s and the time between emitted pulses determines the maximum sensor range $r_{s,max}$.

In order to express the position in the YZ plane of a point p in the ER responsible for a particular bin in the swath, the polar coordinates (r_s, θ_s) are commonly used. The *grazing angle* θ_s can be easily computed as a function of the AUV altitude h, the point altitude h_p and the slant range r_s as follows:

$$\theta_s = \arcsin\left(\frac{h - h_p}{r_s}\right) \tag{1}$$

The AUV altitude h can be obtained either by external sensors, such as a *Doppler Velocity Log* (DVL), or by properly analysing the SSS data, as it will be shown in Section 2.3. The slant range r_s is already available since it is fully defined by the bin. Unfortunately, obtaining the point altitude h_p solely from SSS data [36] is a complex and error prone task and the absence of such altitude information leads to the most serious difficulties in SSS data processing. Accordingly, if bathymetric data is not guaranteed by additional sensors, it is only possible to state that the grazing angle is within an interval defined by the mounting angle θ and the vertical opening α as follows:

$$\theta_s \in \left[\theta - \frac{\alpha}{2}, \theta + \frac{\alpha}{2}\right] \tag{2}$$

This is a large interval, since SSS are built with large α. Because of that, most researchers perform the so called *flat floor assumption*. This means assuming that the ocean floor is flat within the ER and parallel to the XY plane. That is, a common approach is to assume that $h_p = 0$. As a matter of fact, without external bathymetry, this assumption is mandatory in order to make subsequent data processing tractable.

Even though this may seem a hard assumption, there are two aspects to emphasize. First, the flat floor assumption is local. The sensor altitude between the recording of consecutive swaths can change and so the ocean floor is not assumed to be flat along the AUV path. Second, the effects of assuming $h_p = 0$ in Equation (1) decrease as the AUV altitude increases. In this way, the flat floor assumption has almost negligible effects when the AUV navigates at high altitudes $h >> h_p$. An in-depth analysis of the errors introduced by the flat floor assumption is available in [17].

A similar situation arises in the XY plane due to the horizontal opening φ. In this case, as shown in Figure 1, the point p can be anywhere within the arc q. Similarly to what happens in the vertical plane, this means that one or more objects within that arc may be responsible for the received echo intensity. Some studies [17] tackle this problem by fusing data from different swaths to remove the ambiguities. However, this problem is usually neglected by assuming a pencil-like thin beam in this plane. This assumption is reasonable given the small opening φ and the typical speeds at which AUVs move which prevent overlapping in the XY plane between the regions ensonified to grab consecutive swaths.

In order to represent the measurements with respect to a coordinate frame located at the sea floor, the coordinates of each point p in the ER must be properly placed in the sea floor plane. The slant ranges, which are distances from the sea floor to the sensor itself, cannot be directly used and the so called *ground range* r_g is needed. The ground range of a point p is defined as the projection over the y axis of the vector joining the SSS origin of coordinates and the point p. From Figure 1 it is easy to obtain the following expression:

$$r_g = \sqrt{r_s^2 - (h - h_p)^2} \tag{3}$$

Computing the ground range is affected by the same problem that appeared when computing the grazing angle: the altitude of point p is required. Because of that, the flat floor assumption is also commonly applied and h_p is assumed to be zero. Computing the ground ranges is known as *slant range correction*. Our proposal to achieve this goal is provided in Section 3.3.

2.3. Acoustic Image Formation

As stated previously, the SSS is composed of two sensing heads symmetrically mounted on the AUV. Since both sensing heads operate simultaneously, the swaths coming from them are usually joined into a single vector which is called a *full swath*. For the sake of simplicity, the term *swath* will be also used as synonym of full swath whenever there is no ambiguity.

Figure 2 shows an example of a swath gathered with a particular SSS that provides 500 bins. The x axis corresponds to the bin number and, so, the slant range can be computed from it. The first 250 bins have been provided by the port sensing head whilst the last 250 correspond to the starboard sensing head. The y axis represents the received echo intensity normalized to the interval [0,1].

The central region with very low echo intensities, known as *blind zone*, corresponds to time steps for which no sea floor was detected. The blind zone is due to the region below the AUV that has not been ensonified. Thus, the small echo intensity values in this zone are produced by a combination of internal sensor noise and small particles suspended in water.

The first significant echo outside the blind zone is called the *First Bottom Return* (FBR), and it corresponds to the point in the ER closest to the corresponding sensing head. Determining the bin and, thus, the slant range $r_{S,FBR}$ at which the FBR appears is not difficult since the blind zone has almost zero echo intensity values. This is extremely important since the AUV altitude h can be inferred from the slant range of the FBR.

Taking into account that the FBR is due to the portion of the ER closest to the SSS, it is reasonable to assume that the grazing angle of the FBR is $\theta + \frac{\alpha}{2}$. According to the SSS geometry shown in Figure 1, this means that the AUV altitude h is:

$$h = r_{s,FBR} \cdot \sin\left(\theta + \frac{\alpha}{2}\right) + h_{FBR} \tag{4}$$

where h_{FBR}, which is the altitude of the FBR, is the only unknown value. However, if we perform the aforementioned flat floor assumption then $h_{FBR} = 0$ and so the AUV altitude can be computed. Conversely, if the AUV altitude is known by external means, the slant range of the FBR can be computed.

Figure 2 clearly shows another important feature of SSS that has to be taken into account to properly understand and process the data. As it can be observed, the regions surrounding the blind zone have significantly larger intensity values than those far from it. This is particularly visible in the starboard part of this Figure.

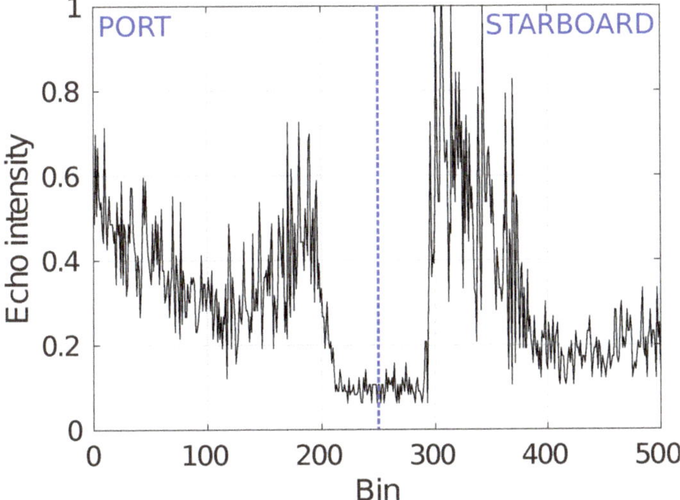

Figure 2. An example of a swath composed of 500 bins. Bins from 0 to 249 come from the port sensing head. Bins from 250 to 499 are provided by the starboard sensing head.

Aside of the reflectivity of the sea floor, which carries information about the environment, there are two other factors that influence the parts of the ER that will produce larger echo intensities. On of these factors in the ensonification pattern, which depends on the sonar configuration and is not homogeneous within the ER. The other factor is the sound attenuation with the travelled distance. In the particular case of SSS, these two factors combine constructively nearby the blind zone, being responsible for the above mentioned larger intensity values. Section 3.2 discusses this issue and proposes a method to reduce the negative effects of this trend in the received echo intensities.

Different swaths are grabbed by the SSS while the AUV moves. By aggregating swaths, an *acoustic image* is built. A common assumption to build these images is that the AUV moved following a straight line usually called *transect*. In this way, building the acoustic image is achieved by simply stacking the swaths one next to the other. This assumption is reasonable, since in most cases AUV equipped with SSS are programmed to follow a straight transect [16], go to surface, turn to the desired direction, submerse again and follow a new straight transect. This study will make this assumption, though some studies exist that make use of the AUV pose to properly account for the exact AUV motion [17].

Figure 3 shows an example of an acoustic image built by putting swaths together and mapping echo intensities to grayscale levels. Dark tonalities correspond to low echo intensities and light tonalities denote high echo intensities. Each column of pixels in the image shows the swath vector obtained from one ping while the AUV was moving from left to right. So, one can think about these images as being built from left to right by adding a new column of pixels at each time step.

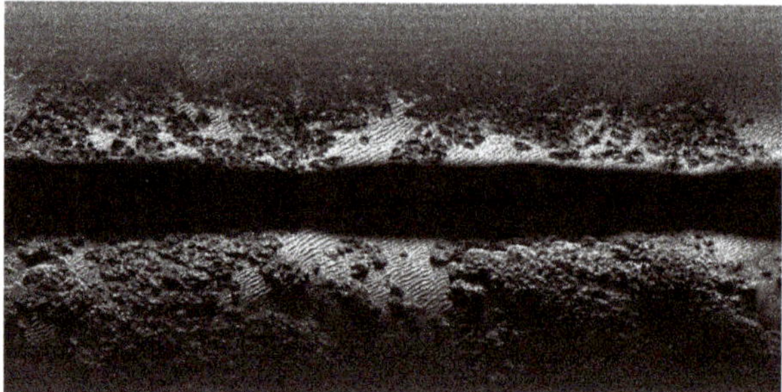

Figure 3. Example of acoustic image. Source: [37].

The central dark strip is the blind zone. Changes in its height reflect changes in the AUV altitude so that the larger the height the larger the altitude, as already shown in Equation (4). The effects of the uneven ensonification and the sound attenuation with distance can also be observed in the bright regions surrounding the blind zone and the overall darkening with distance to the central bin.

3. Data Pre-Processing

3.1. Overview

Given the particularities of the SSS and the acoustic image formation, it is advisable to pre-process the data prior to segmenting the acoustic images. The pre-processing, which has to be performed locally as soon as a new swath arrives to make it possible on-line operation, is performed in two steps each one correcting one of the SSS characteristics mentioned previously.

The first step, called *intensity correction*, tackles the problem of the signal baseline, which is mostly due to the uneven SSS ensonification pattern and the sound attenuation with distance. The second step, called *slant range correction* deals with the problem of the unknown altitudes within the ER. Both steps are described next.

3.2. Intensity Correction

The received echo intensity is the combination of three components. First, the reflectivity of the sea floor. Depending on the characteristics of each point in the ER, different echoes will be produced. Second, the SSS ensonification pattern. Roughly speaking, the emitted sound intensity is much larger nearby the acoustic axis and decreases with the angular distance to the acoustic axis. Third, the sound attenuation with distance. The larger the distance the sound has to travel, the more the energy is lost and, thus, the smaller the received echo intensity. The only component that carries useful information about the environment is the reflectivity of the sea floor. Thus, it is desirable to compensate the other components.

As it can be observed in Figure 1 the SSS acoustic axis intersects the sea floor nearby the FBR, which is the point in the ER closest to the sensor. Thus, under this configuration, both the ensonification pattern and the sound attenuation with distance combine to produce larger echo intensities near the blind zone. This situation in which the ensonification pattern and the sound attenuation reinforce the signal in the same region is common to all SSS configurations, but it is not general to all sonar sensors. For example, in the MSIS described in [38], sound attenuation and ensonification pattern focus on different regions of the ER and the overall effect is that larger echo intensities appear far away from the sensor.

Since the combination of these three components depends on the specific sonar configuration, some researchers deal with it using some sensor and environment dependant heuristics [39]. To the contrary, the proposal in this paper is general, so it can be applied to different sonar configurations, and relies on a well founded theoretical basis. As a matter of fact, the same theory behing our proposal has been successfully applied to both SSS [17] and to MSIS [38].

Our proposal to model the echo intensity $E(p)$ produced by a point $p = (r_s, \theta_s)$ in the ER follows the echo pressure amplitude model by Kleeman and Kuc [40] and is:

$$E(r_s, \theta_s) = \frac{f \cdot a^4}{r_s^2} \left(\frac{2 \cdot J_1\left(\frac{2 \cdot \pi}{\lambda} \cdot a \cdot \sin(\theta_s - \theta)\right)}{\frac{2 \cdot \pi}{\lambda} \cdot a \cdot \sin(\theta_s - \theta)} \right)^2 \tag{5}$$

where f is the emitted pulse frequency, a is the transducer radius, $J_1(\cdot)$ is the Bessel function of the first kind of order 1 and λ is the emitted pulse wavelength. This Equation explicitly accounts for the uneven ensonification pattern, which depends on the angular position of p with respect to the acoustic axis, expressed by the term $\theta_s - \theta$, and the sound attenuation with distance, expressed by the term r_s^2.

The frequency f is usually provided by the SSS manufacturers. The wavelength λ is uniquely related to f given the speed of sound in water, which depends on the water conditions. Even though these conditions may be unknown or mutable, it is reasonable to assume [41] speed of 1560 m/s for SSS operating in sea water or 1480 m/s for SSS operating in freshwater to compute λ given f.

Unfortunately, the transducer radius may not be available. Moreover, the transducer may not even be circular. To alleviate this problem, we propose a method to compute a. If the transducer is actually circular, then the obtained a will represent the radius. To the contrary, if the transducer is not circular, the obtained a will not have a geometric interpretation but still could be used in Equation (5).

Given one sensing head, the blind zone corresponds to grazing angles equal or larger than $\theta + \frac{\alpha}{2}$, as it can be observed in Figure 1. This means that the echo intensity for these angles is zero. In particular, the echo intensity at $\theta + \frac{\alpha}{2}$ is zero. Using this information, that is, the fact that $E(r_s, \theta + \frac{\alpha}{2}) = 0$, to rewrite Equation (5), the following expression is obtained:

$$0 = \frac{f \cdot a^4}{r_s^2} \left(\frac{2 \cdot J_1\left(\frac{2 \cdot \pi}{\lambda} \cdot a \cdot \sin(\theta + \frac{\alpha}{2} - \theta)\right)}{\frac{2 \cdot \pi}{\lambda} \cdot a \cdot \sin(\theta + \frac{\alpha}{2} - \theta)} \right)^2 \tag{6}$$

For this equality to be true, either f or a must be zero, which is physically impossible, or $J_1\left(\frac{2 \cdot \pi}{\lambda} \cdot a \cdot \sin(\theta + \frac{\alpha}{2} - \theta)\right) = 0$. This Bessel function of the first kind J_1 has an infinite number of zeros. According to [40], the first zero corresponds to the boundary of the main ultrasonic lobe whilst subsequent zeros model the boundaries of the secondary side lobes. The energy of these side lobes is usually so small that they are often ignored. Since the first occurring zero of J1 appears at the boundary of the main lobe which is the one modeled by the opening α our proposal is to focus, precisely, on that first zero though the effect of the other zeros could be explored. The first x that makes $J_1(x) = 0$ is approximately $x = 3.8317$ [42]. That is, we can rewrite Equation (6) as follows:

$$3.8317 = \frac{2 \cdot \pi}{\lambda} \cdot a \cdot \sin(\frac{\alpha}{2}) \tag{7}$$

Thanks to that, we can express the transducer radius a as a function of the wavelength and the vertical opening:

$$a = \frac{3.8317 \cdot \lambda}{2 \cdot \pi \cdot \sin\left(\frac{\alpha}{2}\right)} \tag{8}$$

Even though $E(p)$ is the echo intensity produced by point p, this intensity is modulated by the incidence angle. That is, the echo intensity that will reach the SSS depends on the angle at which the sound collides with the sea bottom at point p. This angle is unknown, and cannot be computed unless external bathymetry is available, but it can be approximated by the grazing angle θ_s under the flat floor assumption. Accordingly, if we model the sea floor as a Lambertian surface [36,43] which scatters uniformly the incident energy in all directions, the component of $E(p)$ that reaches the sensor is $E(p) \cdot \cos\theta_s$. Finally, the received echo also depends on the particular acoustic properties of point p. Let us model these properties as $R(p)$, which is called the *reflectivity*.

We can now represent the echo intensity $I(p)$ received by the sensor and echoed by a sea floor point $p = (r_s, \theta_s)$, with the following expression:

$$I(p) = K \cdot R(p) \cdot E(p) \cdot \cos\theta_s \tag{9}$$

where K is a normalization constant. This Equation makes it possible to get the reflectivity, which carries information about the sea floor, as a function of the received echo intensity $I(p)$, which is the actual SSS output, and the sound ensonification intensity $E(p)$, which can be computed using Equation (5):

$$R(p) = \frac{I(p)}{K \cdot E(p) \cdot \cos\theta_s} \tag{10}$$

Since $R(p)$ solely contains information about the sea floor, discarding the uneven ensonification and the attenuation with distance, the intensity correction consists, precisely, on applying Equation (10) to each bin provided by the SSS. An acoustic image built using this corrected data will be referred to as a *intensity corrected* image.

3.3. Slant Range Correction

The term slant range correction refers to the projection of each bin to the corresponding position in the sea floor. This can be achieved by means of Equation (3) if h_p is known. If the point altitude is unknown, then the flat floor assumption $h_p = 0$ can be applied.

However, from an algorithmic point of view, Equation (3) is not practical. Taking into account that the goal of the slant range correction is to create a new swath in which each bin corresponds to a ground range, it is more useful to have an Equation that, given a ground range, provides the corresponding slant range so that it can be used to query the original swath and, thus, provide the echo intensity at that particular ground range. That is, the equation that will be used is derived from Equation (3) and is the following:

$$r_s = \sqrt{r_g^2 + (h - h_p)^2} \tag{11}$$

In order to build the corrected swath there are two additional criteria to decide. The first one is the ground range resolution δ_g. That is, the slant corrected swath will be composed of bins of equal size, and that size needs to be defined. This resolution δ_g can be decided depending on the desired granularity or depending on the mounting angle and the openings, among other factors. However, in general, using the same resolution than the original swath is convenient. This is the approach used in this paper and, thus, $\delta_g = \delta_s$.

The second criteria to decide is related to the fact that, when building the new swath we can evaluate Equation (11) for each ground range r_g corresponding to one specific bin in the corrected swath but the resulting slant range may not correspond to one specific bin in the original swath but lie somewhere between two adjacent bins. In this case, our proposal is to perform linear interpolation.

As a result of this process, a *slant corrected* swath is obtained. The acoustic image obtained by means of these slant corrected swaths is the *slant corrected* image.

3.4. Data Selection

Figure 4 shows the intensity and slant corrected version of the acoustic image in Figure 3. There are two important features to be observed in this image. On the one hand, that the blind zone, outlined in red, carries no information about the sea bottom. On the other hand, that the echo intensity decreases with distance until there is almost no difference between the terrain types. Since this is an intensity corrected image, this means that there is almost no information about the sea bottom from one distance to the central bin onward. Let us call this region the *low contrast zone*.

Accordingly, if our goal is to segment the acoustic image depending on the kind of terrain it depicts, it is desirable to remove both the blind zone, which carries no information, and the low contrast zone, which carries almost no information and can lead to undesired effects when training a NN.

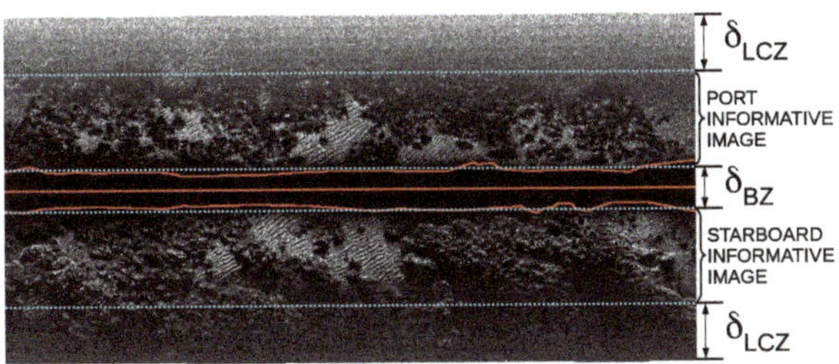

Figure 4. Example of intensity and slant corrected acoustic image. The blind zone as well as the central bins, separating port and starboard, are outlined. The blind and low contrast zones under constant altitude and flat floor assumption are also shown.

The blind zone will be located around central bin of each swath, though the exact size will change from swath to swath because it depends on the AUV altitude, and it will not be symmetrical with respect to the central bin since the FBR can be different for port and starboard sensing heads. These effects can be observed in Figure 4, where it is clear that the blind zone increases or decreases with time and that it is not symmetrical with respect to the central bin.

As for the AUV altitude, this study will assume that the robot navigates at constant altitude. This is not a hard assumption, since most AUV with SSS are programmed to navigate through straight transects at constant altitude. Under this assumption, changes in the blind zone size will only be due to different FBR for port and starboard. However, if we perform the flat floor assumption, which already plays an important role in this study, the FBR should be the same at both sides of the AUV.

Let $I_{N_b \times T}$ denote an intensity and slant corrected acoustic image built from time step 0 to time step $T-1$ by stacking full swaths of N_b bins, the first $\frac{N_b}{2}$ corresponding to the the port sensing head and the last $\frac{N_b}{2}$ corresponding to the starboard sensing head. Under the two aforementioned assumptions, it is possible to define a constant δ_{BZ} so that the blind zone lies within the bins $\frac{N_b-\delta_{BZ}}{2}$ and $\frac{N_b+\delta_{BZ}}{2}$. Similarly, the constant δ_{LCZ} can be defined so that the port low contrast zone lines between the bins 0 and $\delta_{LCZ}-1$ and the starboard low contrast zone is located between $N_b-\delta_{LCZ}$ and N_b-1.

Figure 4 illustrates δ_{LCZ} and δ_{BZ}. As it can be observed, there are two strips in the acoustic image that are considered to carry useful information. One of them, corresponding to the port, comprises bins from δ_{LCZ} to $\frac{N_b-\delta_{BZ}}{2}-1$ and the other one, corresponding to the starboard, lies within bins $\frac{N_b+\delta_{BZ}}{2}+1$ and $N_b-\delta_{LCZ}-1$. Let the bins within these intervals be called the *informative bins*.

The exact values of δ_{BZ} and δ_{LCZ} depend on the specific sensor being used, the average altitude and the environment and will be discussed in Section 5.1.

From now on, only the informative bins of the intensity and slant corrected image will be considered, leading to two informative strips of data per acoustic image. Let the two sets of informative bins within a swath be referred to as *informative swaths* and let the term *informative image* denote the image built by stacking informative swaths, so that two informative images are available for each original acoustic image.

4. Data Segmentation

4.1. Overview

The main goal of this study is to segment acoustic images in order to detect the existing terrain types. The segmentation is mainly meant to be used to provide loop candidates to a SLAM system, though many other applications can benefit from it. Because of that, the acoustic image has to be segmented on-line, ideally swath by swath. That is why standard image segmentation approaches, which require full images instead of swaths (i.e., columns of pixels), cannot be used or have to be adapted to achieve this goal.

Since the swaths provided by the SSS are affected by several sources of error, such as uneven ensonification patterns or geometric distortions, the segmentation will be performed over the informative swaths as described in Section 3.4. That is, the SSS output will be first intensity and slant corrected and then the blind and the low contrast zones will be removed. The remaining bins are those that will be used to feed the data segmentation process.

Our proposal to perform on-line SSS segmentation is based on a CNN. Roughly speaking, a sliding window of the most recently gathered informative swaths will be used to feed the CNN. By using a CNN and a sliding window, each informative swath will be segmented more than once. Thus, a method to combine several segmentations of a single swath is required.

The proposed CNN architecture, as well as its training and usage, are presented in Section 4.2. The method to combine the different proposed segmentations within the sliding window and build a consistent segmentation of the environment is described in Section 4.3.

4.2. The Neural Network

The proposed NN is a fully CNN that follows the encoder-decoder architecture shown in Figure 5, since this kind of architectures define a good compromise between quality and speed to segment small images [44]. Hyperparameters such as the number of layers and the convolutions, pooling and upsampling masks shapes and sizes have been tuned by means of a grid search method.

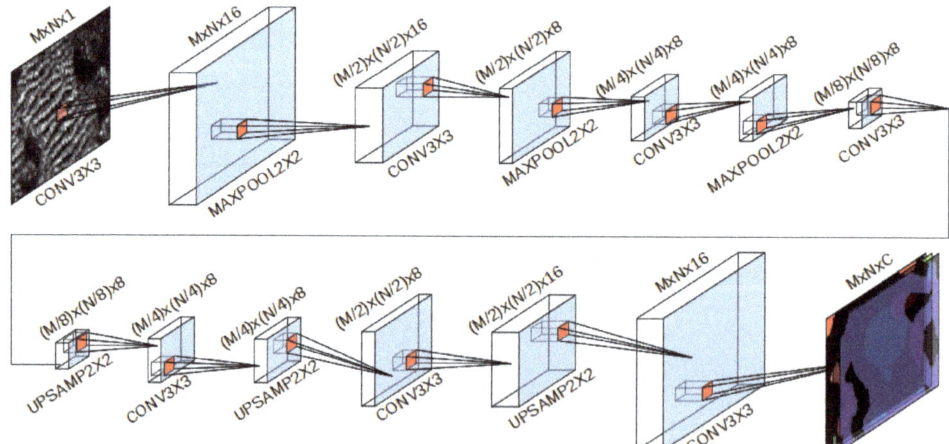

Figure 5. The Neural Network architecture.

The input of this NN is a set of consecutively gathered informative swaths, which constitute a patch of the informative image. The patches, which come from a sliding window over one informative image, have to be joined back to build a segmented informative image. Also, joining the port and starboard segmented informative images to build a full segmented acoustic image is necessary. Both tasks have to be performed externally to the NN as it will be described in Section 4.3.

The encoder part of the NN reduces the dimensionality of the input patch to the so called *latent space* by means of a set of convolutional and max-pooling layers. The latent space is meant to learn the most important features of the input patch, so it can be expanded back to the original size by the decoder.

The decoder part of the NN is composed of couples of upsampling and convolutional layers, each one increasing the dimensionality of the previous one until the original size is reached. The last layer is built using a soft-max activation function, so that each of the C layers expresses the probability of each bin to belong to one of the C classes.

The specific value of C depends on the specific application where the NN is to be deployed and the environment particularities. In our case, as it will be described in Section 5, we used $C = 3$ meaning that three classes, namely rock, sand and others, will be detected, though our proposal is neither targeted nor constrained to any specific number of classes.

4.2.1. Training

In order to train the NN, pairs of informative images and the corresponding ground truth are required. The ground truth images are matrices of the same size that the corresponding informative images where each cell holds a value between 0 and $C - 1$ stating the class to which the corresponding bin in the informative image belongs. The ground truth has to be built manually, by hand-labelling each of the bins.

For each of these pairs of informative and ground truth images, the swaths separated pS time steps between them are selected. In order to reduce overfitting, the selected swaths can be randomly shuffled to remove any sense of order between them.

Then, one informative patch is built for each of these swaths by using the pM preceding and the pM subsequent swaths. That is, one informative patch is composed of $2 \cdot pM + 1$ swaths and it is guaranteed that the swath at the center of the patch is $k \cdot pS, k \geq 1$ swaths away from any other central swath. The corresponding patch in the ground truth image is also selected. The informative patch is used to feed the NN and the ground truth patch is compared to the NN output to provide feedback during training by means of a cross-entropy loss function. Figure 6 illustrates this idea.

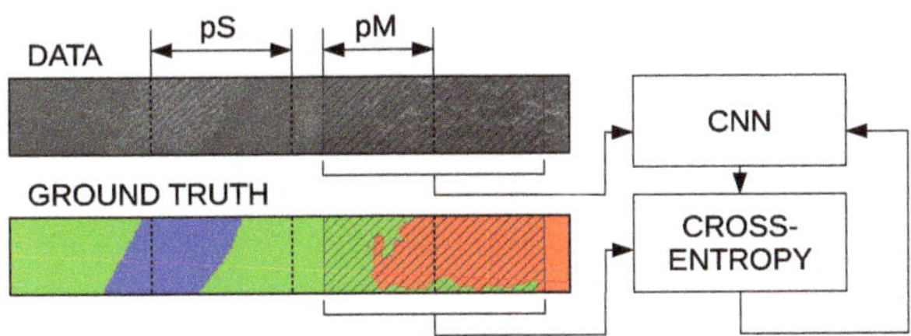

Figure 6. The training process.

Thanks to the *patch separation* (pS) and the *patch margin* (pM) it is possible to define several NN training strategies. For example, a sliding window over the whole informative image can be used to train the NN by simply setting $pS = 1$. Also, strictly non overlapping informative patches can be used just by setting $pS = 2 \cdot pM + 1$.

The specific values of pS and pM, which should be selected taking into account the compromise between training time, training quality and possible overfitting, will be experimentally assessed in Section 5.3.

4.2.2. On-Line Usage

As shown in Figure 5, the input of the NN is an informative patch of M rows and N columns, M being the number of bins in each informative swath and N being $2 \cdot pM + 1$.

Our proposal to perform on-line segmentation of the informative images, which is summarized in Figure 7, is as follows. First, every Δt time steps, an informative patch is built containing the most recent N informative swaths. Then, the informative patch is used to feed the NN and the C output probability layers are obtained. Since each cell in each of the C output layers represent the probability of the corresponding bin to be of one class or another, this means that every Δt time steps we have a segmented patch.

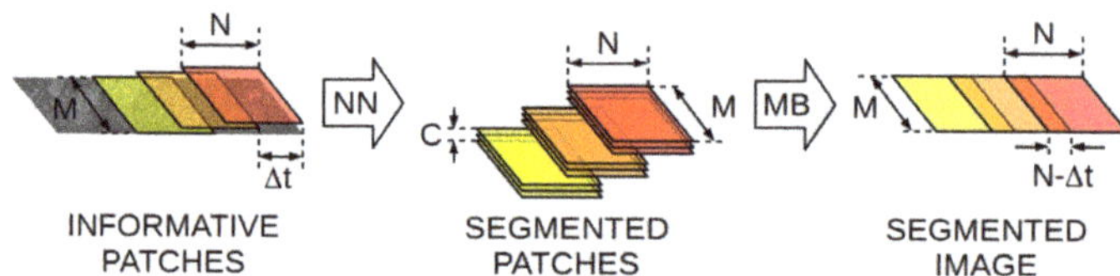

Figure 7. On-line usage of the NN and the Map Building (MB).

If $\Delta t < N$, consecutively segmented patches will have $N - \Delta t$ swaths in common. In other words, each swath will be segmented more than once. Our proposal to combine these multiple segmentations per swath will be discussed in Section 4.3. Using $\Delta t > N$ is not advisable since will produce gaps in the segmented acoustic image.

Low values for Δt lead to low latency segmentation. For example, $\Delta t = 1$ means that every new swath is segmented as soon as it is gathered. However, low values of Δt also lead to larger computational demand. So, deciding the specific value for this parameter depends on a compromise between latency and computational burden. An experimental assessment on this regard will be performed in Section 5.3.

4.3. Map Building

The goal of the *Map Building* (MB) is to join the segmented patches provided by the NN into a single, consistent, *segmented image*. If $\Delta t < N$ there will be overlapping between segmented patches and so the MB has to properly combine them, as illustrated in Figure 7. In this paper, two different methods to build the segmented image are presented: the *single-class method* (SCM) and the *multi-class method* (MCM).

The SCM assigns a single label to each bin in the segmented image stating its class. The process begins by assigning a single label to each bin in each segmented patch. The assigned label is the one corresponding to the class with the highest probability. If $\Delta t = N$ each swath took part in one and only one of the segmented patches and, so, the classes assigned to the segmented patches can be directly placed into the segmented image. However, if $\Delta t < N$ each swath was used to build more than one of the informative patches classified by the NN. In this case, the label assigned to each bin in the segmented image is the majority class of the corresponding bins in all the involved patches. As a result of this process, each bin in the segmented image is assigned to one and only one class.

The MCM keeps the same structure of the segmented patches in the segmented image. That is, the segmented image will be composed of C layers, each one stating the probability of each bin to belong to each of the C classes. If $\Delta t = N$ the C probability layers present in each segmented patch can be directly placed into the segmented image. If $\Delta t < N$, the average probability of each class in all the overlapping patches is placed in the corresponding positions of the segmented image. Finally, all the probabilities are normalized to sum one.

As an example, Figure 8 shows the results of using SCM and MCM to build segmented images being the number of classes $C = 3$. In order to represent the classification, the three primary colors red, green and blue have been assigned to each class. As stated previously, $C = 3$ is the specific number of classes that will be used in the experiments, though other values could be used depending on the target application and the sea floor structure.

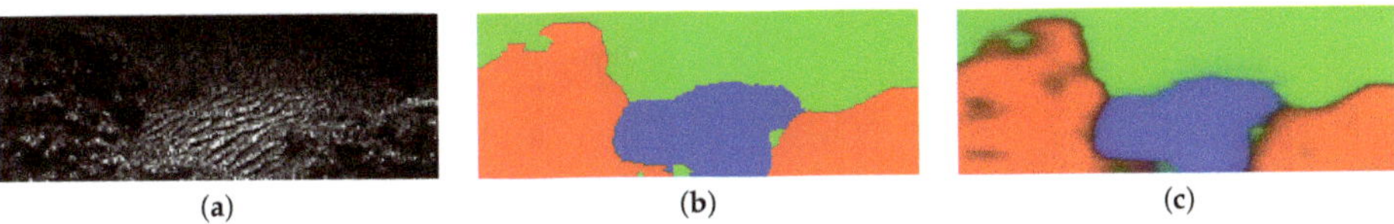

(**a**) (**b**) (**c**)

Figure 8. Example of (**a**) a set of informative swaths and the corresponding segmented images using (**b**) SCM and (**c**) MCM.

Figure 8b shows the output of SCM when used to build the segmented image from the informative swaths in Figure 8a. As it can be observed, a single label is assigned to each bin, resulting in a clear red, green or blue color per pixel.

Figure 8c shows the corresponding MCM segmented image. In this case the probability layers are kept for each patch and combined to build the segmented image. As a result, the depicted colors are a combination of red, green and blue depending on the probability of the corresponding class. Because of that, it is easy to visualize uncertainties in the class contours and also appreciate some details that are not visible in the SCM, such as the small green gaps within the red blob in the left part of the image.

5. Experimental Results

5.1. Overview

The data used to perform the experiments has been obtained by an EcoMapper AUV equipped with an Imagenex SportScan SSS, whose main parameters are summarized in Table 1 using the notation presented in Section 2.

Table 1. Parameters of the Imagenex SportScan SSS used in this paper.

α	30°
φ	3°
θ	20°
f	800 KHz
λ	1.95 mm
$r_{s,max}$	30 m
δ_s	0.12 m
Bins per swath	250 port, 250 starboard
Sampling frequency	10 swath/s

The AUV mission consisted of a sweeping trajectory along more than 4 Km in Port de Sóller (Mallorca, Spain). During the straight transects the AUV was underwater gathering SSS data at an approximate altitude of 5 m. At the end of every straight transect the AUV stopped recording SSS, surfaced, changed to the new orientation while correcting its pose estimate using GPS and submersed again to gather data along a new straight transect. Accordingly, the gathered SSS data correspond to straight transects at almost constant altitude.

The AUV was also equipped with a *Doppler Velocity Log* (DVL) sensor, providing instantaneous speed information as well as precise altitude and heading measurements. By combining DVL when the AUV was underwater and GPS when the AUV surfaced, the trajectory followed by the AUV can be computed [17]. This trajectory, shown in Figure 9 overlaid to a Google Maps satellite view, illustrates the mission performed by the AUV while gathering the data used in this paper.

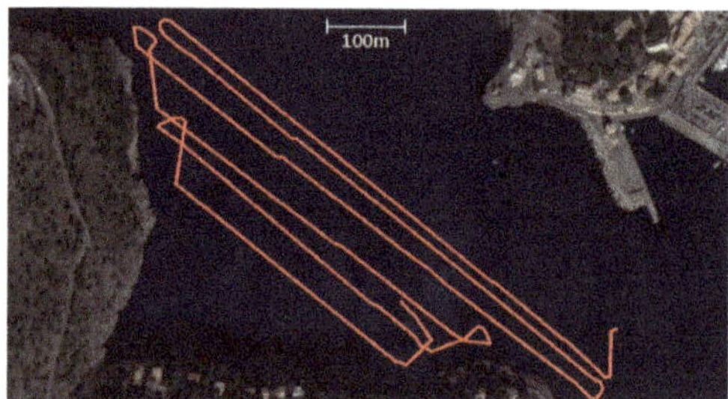

Figure 9. Trajectory followed by the AUV.

The short transects in which the AUV just submersed and surfaced with almost no motion at constant altitude have been removed from the data used in this paper, leading to a dataset composed of five transects and, thus, ten informative images. These transects involve a total of 22438 swaths, distributed as shown in Table 2.

In order to test and evaluate our proposal, three different classes have been defined. Two of these classes actually correspond to geological structures: *rock*, exemplified in Figure 10a, and *rippled sand* or sand for short, exemplified in Figure 10b. The third class is called *others* and, even though it mostly corresponds to sand, it actually represents all the data whose texture is not sufficient for a human to properly identify the true sea floor structure. Figure 10c,d show two examples of this class.

Table 2. Dataset specification. The number of informative bins of each class and the corresponding percentage, within parenthesis, are provided.

Transect	Swaths	Rock	Sand	Other
1	5764	125294 (13.095%)	26464 (2.766%)	805066 (84.139%)
2	6800	107706 (9.542%)	133665 (11.841%)	887429 (78.617%)
3	3825	253029 (39.850%)	111411 (17.546%)	270510 (42.603%)
4	3517	215148 (36.852%)	70604 (12.093%)	298070 (51.055%)
5	2532	136762 (32.538%)	38650 (9.196%)	244900 (58.266%)
GLOBAL	22438	837939 (22.497%)	380794 (10.223%)	2505975 (67.280%)

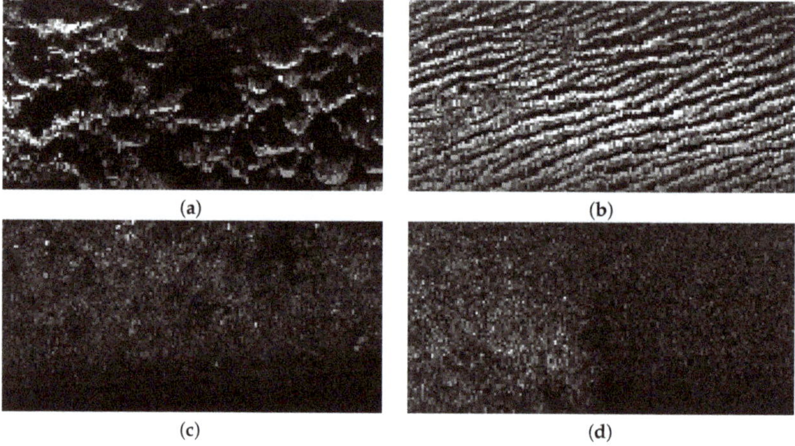

Figure 10. Examples of the three considered classes: (**a**) rock, (**b**) rippled sand and (**c**,**d**) other.

A ground truth has been constructed, both to train our NN and to test it, by hand-labelling each bin in each of the acoustic images with the corresponding class.

These classes are not equally distributed along the transects. As it can be expected, the class *others* is the most frequent one, representing the 67.28% of the whole dataset. The other classes represent the 22.497% (rock) and the 10.223% (sand). The percentage of each class in each of the transects is detailed in Table 2. Since classes are unbalanced, the quality measures have to take into account that particularity.

Next, the experiments are presented and discussed. First, the system is calibrated in Section 5.2. Afterwards, a complete set of experiments and quantitative results is shown in Section 5.3. Finally, some qualitative results are provided in Section 5.4.

5.2. System Parametrization

The SSS data has been processed as described in Section 3. To this end, the ensonification pattern, modelled by Equation (5), has been computed for each emitted ping and, thus, for each gathered swath using the parametrization shown in Table 1. Figure 11a depicts the obtained values along a short transect. Changes along track are due to changes in altitude. As it can be observed, the ensonification pattern clearly reflects the two sensing heads and the blind zone. It also illustrates the two peaks showing the parts of the ER that will be ensonified with more energy. Figure 11b shows the same values in a 2D plane where the color intensity illustrates the ensonification intensity.

Figure 11c shows the SSS data gathered along the same short transect used to build the ensonification pattern. As it can be observed, the regions close to the blind zone are responsible for very large echo intensities, reaching a condition close to saturation and making it difficult to distinguish objects within these regions. This effect can be also observed in Figure 11d, which shows the same data in the bin-swath plane using color intensity to represent the echo intensity.

The ensonification pattern is used to correct the raw SSS data. Thanks to that, the echo intensity is homogenized, desaturating the regions close to the blind zone and thus emphasizing the existing objects in the acoustic image. Figure 11e shows the result of applying the ensonification pattern to the raw SSS data using Equation (10) and performing the slant range correction by means of Equation (11). The same data projected to a 2D plane is depicted in Figure 11f.

As it can be observed, the objects close to the blind zone are more distinguishable from the background that in the original data, revealing some small details that were not appreciable in the raw SSS swaths. This process can be seen as a physics based contrast enhancement that leads to an homogeneous contrast almost independently of the bin location, thus helping the operation of segmentation algorithms.

By observing the examples in Figure 11 it can be seen not only the blind zone but also that, from a certain bin onward, both on port and on starboard, the echo intensity is so small that is is difficult to clearly ascertain the structure of the ocean floor, even in the intensity corrected version. This is particularly clear in the starboard part of Figure 11d,f. These are, precisely, the low contrast zones mentioned in Section 3.4.

As mentioned in that section, to reduce the problems that the non informative blind and low contrast zones would induce in the subsequent segmentation, they are removed. To this end, the parameters δ_{BZ} and δ_{LCZ} have been defined. In order to determine these parameters, we proceeded as follows.

First, we computed the average FBR using Equation (4), the average AUV altitude and performing the flat floor assumption. This obtained average FBR is 25 bins (both on port and starboard) and is directly related to δ_{BZ}. As a matter of fact, according to Figure 4, δ_{BZ} should twice the FBR. Thus, we determined in this way that $\delta_{BZ} = 50$.

Since the low contrast zone is mainly due to the low ensonification intensity for large distances, we used Equation (5) to determine δ_{LCZ}. More specifically, given that $\delta_{BZ} = 50$, we have searched the δ_{LCZ} that keeps the 90% of the ensonification intensity within each informative image. By using this procedure, we have found that $\delta_{LCZ} = 142$.

Taking into account that the SSS used in the experiments provides 250 bins per sensing head, this means that each informative image, as defined in Section 3.4 and illustrated in Figure 4, is composed of 83 bins. This an approximation based on the assumption of a flat floor and a constant navigation altitude. Even though computing δ_{BZ} and δ_{LCZ} on-line using instantaneous altitude measurements may seem a better option, that approach would lead to changes in size of the informative image which would be problematic in further segmentation steps. That is why this study uses the constant δ_{BZ} and δ_{LCZ} approximation.

Finally, the value of the patch margin pM, presented in Section 4.2.1, has been set to $pM = 41$ so that the number of swaths in patch, which is $2 \cdot pM + 1$, equals the number of bins, which is 83. Thanks to this, the NN will be fed with square patches.

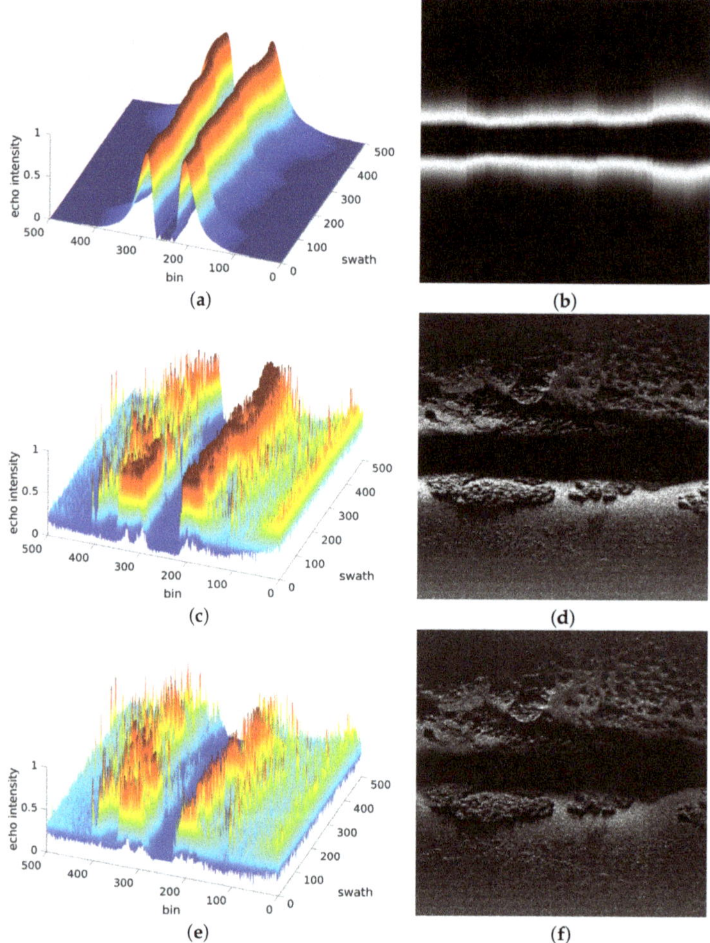

Figure 11. Example of data processing. (**a**,**b**): Modelled echo intensity $E(p)$ according to Equation (5). (**c**,**d**): Raw SSS data. (**e**,**f**): Intensity and slant corrected acoustic image.

5.3. Quantitative Results

After tuning δ_{BZ}, δ_{LCZ} and pM we conducted some experiments to quantitatively evaluate our proposal and the effect of pS and Δt. As explained in Section 4.2.1, the patch separation pS defines the number of swaths between the centers of the patches used to feed the NN during training. In this way, a value of $pS = 1$ means that a sliding window over the whole informative images is used to train the system and $pS = 2 \cdot pM + 1 = N$ uses strictly non overlapping patches to train the NN. Values larger than N are not considered since that means that some input data is discarded. Figure 6 illustrates the meaning of pS. Independently of the value of pS, the order in which the patches are used to feed the NN is randomized in order to prevent overfitting.

The meaning of Δt, which also lies in the interval [1,N], is similar to the one of pS but it refers to the separation between patches during the on-line usage of the NN. Its was explained in Section 4.2.2.

The tested values of pS are 1, $\frac{N-1}{2}$ and N. The tested values for Δt are also 1, $\frac{N-1}{2}$ and N. In this way, we explore the effects of using small, medium and large values for both parameters. Given that in our case $N = 83$, this means that the explored values are 1, 41 and 83. Both the single class (SCM) and the multi class (MCM) methods have been evaluated using all the combinations of parameters mentioned before. This leads to eighteen tested configurations, nine being for SCM and nine for MCM.

For each of the mentioned eighteen combinations, a K-Fold cross validation with $K = 5$ has been performed. To evaluate the quality, the resulting segmented image has been compared to the ground truth and the confusion matrix has been constructed. In the case of MCM, the most probable class for each bin in the resulting segmented image has been used to do the comparison.

Let the components of the confusion matrix be named $N_{x,y}$, so that $N_{x,y}$ denotes the number of bins predicted to be of class x that actually are of class y, where x and y can be 0, 1 or 2, denoting the classes *rock*, *sand* and *others* respectively. Thus, the correct classifications are those where $x = y$.

It is important to emphasize that the decision of a classification being correct or not is performed by comparing the classification itself to a hand labelled ground truth. This ground truth is, by definition, imperfect since it can be subject to human interpretation. Also, some regions may be difficult to classify even for a human, especially in the boundaries between classes and some subjective decisions have to be made in these cases. Thus, the presented results can be slightly influenced by these imperfections in the ground truth.

Confusion matrices are a useful tool to quantify and visualize how the segmentation errors are distributed among classes and what classes are more likely to be wrongly classified as another one. In order to provide a clear representation, these matrices are often normalized according to two methods. It is important to emphasize that these methods actually provide the same information but from a different point of view.

The first method is the column-wise normalization, which scales the columns down to sum one. Since columns depict the true classes, column-wise normalization means that the value in row r and column c represents the ratio of bins whose true class is c that have been classified as class r. Thus, this kind of normalization emphasizes the distribution of classes in which the bins of a specific class have been classified.

The second method is the row-wise normalization. In this case, the rows are scaled down to sum one. Rows representing the predicted classes, this format means that the value in row r and column c represents the ratio of bins classified as class r that actually are of class c. Thus, this normalization approach shows the ratio of each of the true classes given the bins predicted to belong to one specific class.

Since eighteen different configuration have been tested, considering these two normalization methods leads to a total of 36 confusion matrices. All these matrices are available at https://github.com/aburguera/NNSSS/tree/master/RESULTS. A summary is provided in Tables 3 and 4.

More specifically, Table 3a shows the confusion matrix corresponding to all the configurations of pS and Δt using SCM and normalized column-wise. It can be observed how the largest values appear in the diagonal, meaning that the ratio of bins correctly identified is the largest one. It can also be observed how classes are confused among themselves. For example, the matrix shows that the 16.25% and the 4.57% of the rocks have been classified as sand and other respectively, thus emphasizing that rocks are misclassified as sand about four times more than they are confused with other.

Table 3b shows the SCM row-wise normalized confusion matrix. It can be observed, for example, that the 88.03% of the bins classified as rock were actually rocks and that the 10.22% and the 1.74% were actually sand and other respectively. Thus, given one bin wrongly classified as rock it is about ten times more likely that it actually is sand than other.

Table 4a,b show the MCM confusion matrices normalized column-wise and row-wise respectively. By comparing them to their SCM counterparts it can be observed that the differences are really small, though suggesting that only minor improvements arise from the use of MCM.

Table 3. Confusion matrices for SCM normalized (**a**) column-wise (**b**) row-wise .

(a)

True / Pred.	Rock	Sand	Other
Rock	0.7917	0.0347	0.0428
Sand	0.1625	0.9409	0.1308
Other	0.0457	0.0242	0.8263

(b)

True / Pred.	Rock	Sand	Other
Rock	0.8803	0.1022	0.0174
Sand	0.0602	0.9221	0.0177
Other	0.1111	0.1558	0.7331

Table 4. Confusion matrices for MCM normalized (**a**) column-wise (**b**) row-wise.

(a)

True / Pred.	Rock	Sand	Other
Rock	0.8015	0.0367	0.0461
Sand	0.1554	0.9399	0.1358
Other	0.0430	0.0233	0.8180

(b)

True / Pred.	Rock	Sand	Other
Rock	0.8721	0.1085	0.0193
Sand	0.0563	0.9247	0.0189
Other	0.1021	0.1502	0.7476

Using the raw confusion matrices, different quality indicators have been computed. The first one is the accuracy A, defined as the ratio of correctly classified bins with respect to the total number of bins being classified:

$$A = \frac{\sum_{i=0}^{2} N_{i,i}}{\sum_{i=0}^{2}\sum_{j=0}^{2} N_{i,j}} \tag{12}$$

The obtained results for SCM and MCM are shown in Table 5. There are no significant differences between the single class and the multi class approaches, independently of the values of pS and Δt.

Table 5. Accuracy results for SCM and MCM.

ACCURACY (SCM)			
pS \ Δt	1	$\frac{N-1}{2}$	N
1	0.9100	0.9020	0.9020
$\frac{N-1}{2}$	0.8990	0.8950	0.8930
N	0.8830	0.8770	0.8790
ACCURACY (MCM)			
pS \ Δt	1	$\frac{N-1}{2}$	N
1	0.9100	0.9070	0.9020
$\frac{N-1}{2}$	0.8990	0.8970	0.8930
N	0.8840	0.8820	0.8790

Also, it can be observed how, overall, the accuracy decreases as the value of pS or Δt increases. In all cases, however, the accuracy is really high, ranging between an 87.7% in the worst case (SCM, $pS = N$ and $\Delta t = \frac{N-1}{2}$) to a 91.0% in the best case (SCM and MCM, $pS = \Delta t = 1$).

The results also show that Δt has less influence in the resulting accuracy that pS. This fact is particularly interesting because small values of pS or Δt lead to larger computational requirements, as it will be shown later. Since pS is only used during training, a small value of pS will not influence the on-line usage of our system and a large value could be used for Δt, allowing a fast segmentation without compromising the quality.

Since our proposal is multi-class, let us evaluate its performance for each of the three proposed classes. To this end, the multi-class versions of the precision, recall, fall-out and F1-score indicators will be used.

The precision P_i of the class i is defined as the ratio between the number of bins correctly classified as being of class i and the total number of bins classified as class i, both correct and incorrect:

$$P_i = \frac{N_{i,i}}{\sum_{j=0}^{2} N_{i,j}} \tag{13}$$

The recall R_i, also known as sensitivity, of the class i is the ratio between the number of bins correctly classified as being of class i with respect to the total number of bins that actually are of class i, independently of how they have been classified:

$$R_i = \frac{N_{i,i}}{\sum_{j=0}^{2} N_{j,i}} \tag{14}$$

The fall-out F_i of class i is the ratio between the number of bins incorrectly classified as being of class i and the number the number of bins which are not of class i independently of how they have been classified.

$$F_i = \frac{\left(\sum_{j=0}^{2} N_{i,j}\right) - Ni,i}{\left(\sum_{j=0}^{2}\sum_{k=0}^{2} N_{j,k}\right) - \sum_{j=0}^{2} N_{j,i}} \tag{15}$$

Finally, the F1-Score $F1_i$ is the harmonic mean of the precision and the recall and is computed as follows:

$$F1_i = 2 \cdot \frac{P_i \cdot R_i}{P_i + R_i} \tag{16}$$

Overall, the precision measures how reliable are the segmentation results for each class. It can be seen as the probability of a bin classified in one particular class to actually be of that particular class. The recall measures how complete are the segmentation results for each class, since it measures the fraction of the existing bins in that class that have been properly detected. The fall-out is a measure of the errors when classifying each class. Finally, the F1-Score, which combines precision and recall, is said to be a particularly good indicator when it comes to unbalanced datasets, which is likely to happen in underwater scenarios such as the one where our dataset has been collected (see Table 2).

Accordingly, a good segmentation would result on large values ($\simeq 1$) of P_i, R_i and $F1_i$ and small values ($\simeq 0$) of F_i, and discrepancies between the indicators would provide valuable information.

Table 6 shows the obtained results when using SCM. Results are consistent between the indicators and they show how quality tends to decrease as pS and Δt increase, though the effects of Δt deserve further analysis.

Table 6. Precision, recall, fall-out and F1-Score results when using SCM.

Precision

pS \ Δt	1			$\frac{N-1}{2}$			N		
	Rock	Sand	Other	Rock	Sand	Other	Rock	Sand	Other
1	0.8860	0.9350	0.7960	0.9050	0.9250	0.7430	0.8720	0.9320	0.7780
$\frac{N-1}{2}$	0.8560	0.9370	0.7430	0.8770	0.9310	0.6960	0.8430	0.9360	0.7240
N	0.8950	0.9030	0.7290	0.9080	0.8970	0.6740	0.8810	0.9030	0.7150

Recall

pS \ Δt	1			$\frac{N-1}{2}$			N		
	Rock	Sand	Other	Rock	Sand	Other	Rock	Sand	Other
1	0.8300	0.9420	0.8850	0.7940	0.9430	0.9030	0.8190	0.9360	0.8710
$\frac{N-1}{2}$	0.8270	0.9340	0.8190	0.8000	0.9360	0.8450	0.8170	0.9290	0.8190
N	0.7630	0.9510	0.7640	0.7340	0.9510	0.7910	0.7570	0.9470	0.7530

Fall-Out

pS \ Δt	1			$\frac{N-1}{2}$			N		
	Rock	Sand	Other	Rock	Sand	Other	Rock	Sand	Other
1	0.0340	0.1310	0.0230	0.0280	0.1480	0.0290	0.0380	0.1390	0.0250
$\frac{N-1}{2}$	0.0420	0.1310	0.0290	0.0370	0.1410	0.0340	0.0460	0.1350	0.0310
N	0.0320	0.1810	0.0310	0.0280	0.1890	0.0370	0.0360	0.1830	0.0330

F1-Score

pS \ Δt	1			$\frac{N-1}{2}$			N		
	Rock	Sand	Other	Rock	Sand	Other	Rock	Sand	Other
1	0.8570	0.9390	0.8380	0.8460	0.9340	0.8160	0.8450	0.9340	0.8220
$\frac{N-1}{2}$	0.8410	0.9350	0.7790	0.8360	0.9330	0.7630	0.8300	0.9320	0.7690
N	0.8240	0.9270	0.7460	0.8120	0.9230	0.7280	0.8140	0.9240	0.7340

Also, these results make it possible to observe the differences between classes. More specifically, the best precision, recall and F1-score appear with the sand class. This means that sand is reliably detected, with precisions larger than 90% in all cases except one ($pS = N$ and $\Delta t = \frac{N-1}{2}$ with $P_1 = 89.7\%$), and almost completely detected, with recalls larger than 92% in all cases and close to 95% in most of the cases. This is a reasonable result, since the class sand corresponds to rippled sand, which has the characteristic pattern shown in Figure 10b, whilst the other classes encompass different textures. Nevertheless, both the class rock and the class others also lead to large precisions, recalls and F1-Scores.

When it comes to fall-out, sand is the class responsible for the worst results. Rock and others have fall-outs below 5% in all cases but sand depicts fall-outs ranging from 13% to 18%. This is likely to be due to the particular shapes in which the sand regions appear in the sea bottom. Whereas rocks and others appear in large regions, usually filling several consecutive swaths both on port and starboard, sand tends to be present in small banks. This means that the perimeter of the sand regions is large within the dataset in comparison to the perimeter of the other classes. Since the perimeter is the most difficult region to segment, even for a human when building the ground truth, the effects of these errors is more noticeable for the sand class.

Figure 12 summarizes the obtained F1-Scores and facilitates the analysis of the effects of Δt. In particular, it can be observed how, independently of pS, the differences in quality between $\Delta t = \frac{N-1}{2}$ and $\Delta t = N$ are really small and, in some cases, using $\Delta t = N$ seems to lead to a small improvement. This suggests that values of Δt within the interval $[\frac{N-1}{2}, N]$ barely influence the segmentation quality.

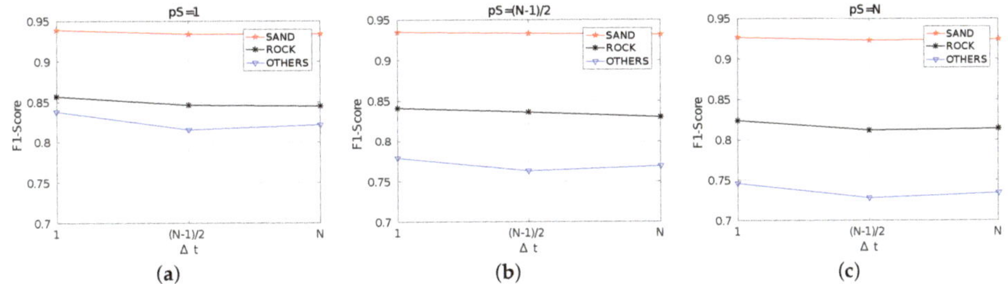

Figure 12. F1-Scores for SCM training with (**a**) $pS = 1$, (**b**) $pS = \frac{N-1}{2}$ and (**c**) $pS = N$.

By comparing Figure 12a–c it can be observed that the results are clearly affected by pS, getting worse as pS increases. It can also be observed how the quality differences between classes increases with pS. Whereas the F1-Score of the sand class remains almost unchanged with pS, the F1-Score of the class others is significantly affected. This suggests that using $pS = 1$ seems to be the best choice.

The results corresponding to MCM are shown in Table 7. These results are numerically similar to those obtained with SCM, and similar trends and patterns can be observed. Thus, the same analysis performed for SCM can be applied here.

Table 7. Precision, recall, fall-out and F1-Score results when using MCM. The gray cells denote the configurations under which MCM surpasses SCM.

Precision

pS \ Δt	1			$\frac{N-1}{2}$			N		
	Rock	**Sand**	**Other**	**Rock**	**Sand**	**Other**	**Rock**	**Sand**	**Other**
1	0.8870	0.9360	0.7950	0.8780	0.9340	0.7890	0.8720	0.9320	0.7780
$\frac{N-1}{2}$	0.8550	0.9380	0.7400	0.8520	0.9370	0.7370	0.8430	0.9360	0.7240
N	0.8930	0.9040	0.7280	0.8880	0.9040	0.7230	0.8810	0.9030	0.7150

Recall

pS \ Δt	1			$\frac{N-1}{2}$			N		
	Rock	**Sand**	**Other**	**Rock**	**Sand**	**Other**	**Rock**	**Sand**	**Other**
1	0.8300	0.9420	0.8870	0.8240	0.9400	0.8780	0.8190	0.9360	0.8710
$\frac{N-1}{2}$	0.8280	0.9340	0.8210	0.8260	0.9320	0.8190	0.8170	0.9290	0.8190
N	0.7650	0.9510	0.7650	0.7620	0.9490	0.7590	0.7570	0.9470	0.7530

Fall-Out

pS \ Δt	1			$\frac{N-1}{2}$			N		
	Rock	**Sand**	**Other**	**Rock**	**Sand**	**Other**	**Rock**	**Sand**	**Other**
1	0.0330	0.1310	0.0230	0.0360	0.1340	0.0240	0.0380	0.1390	0.0250
$\frac{N-1}{2}$	0.0420	0.1300	0.0290	0.0430	0.1320	0.0300	0.0460	0.1350	0.0310
N	0.0330	0.1790	0.0310	0.0340	0.1800	0.0320	0.0360	0.1830	0.0330

F1-Score

pS \ Δt	1			$\frac{N-1}{2}$			N		
	Rock	**Sand**	**Other**	**Rock**	**Sand**	**Other**	**Rock**	**Sand**	**Other**
1	0.8570	0.9390	0.8390	0.8500	0.9370	0.8310	0.8450	0.9340	0.8220
$\frac{N-1}{2}$	0.8410	0.9360	0.7780	0.8380	0.9350	0.7760	0.8300	0.9320	0.7690
N	0.8240	0.9270	0.7460	0.8200	0.9260	0.7400	0.8140	0.9240	0.7340

However, interesting conclusions arise when observing the cases in which MCM surpasses SCM. These situations are those shown in gray cells in Table 7, which mark the cases in which precision, recall, fallout and F1-Score values are larger for MCM and fall-out is smaller. The first aspect to emphasize is that differences, in all cases, are small. So, even though MCM improves SCM in some cases and leads to worse results in some others, the overall quality is almost the same.

The second aspect to emphasize, and probably the most relevant, is related to how the cases in which MCM improves SCM are distributed. For the sake of simplicity, let us focus on the F1-Score, though similar patterns appear with the other indicators.

As it can be observed, the improvements mostly depend on Δt and are almost uncorrelated wit pS, which is reasonable since the use SCM or MCM has no effect during training. It can also be observed that for $\Delta t = N$, MCM never surpasses SCM. However, this does not mean that MCM is worse in this case since the scores are exactly the same, within the working precision, for SCM and MCM. This is also reasonable, because $\Delta t = N$ means that the segmented patches do not overlap. Since the differences between SCM and MCM are the way in which overlapping regions are fused, no differences should appear in this case. It is important to emphasize that results are exactly the same in that case because the same trained model was used both for SCM and MCM since the data fusion method does not affect training.

Thus, the two interesting cases are $\Delta t = 1$ and $\Delta t = \frac{N-1}{2}$. For $\Delta t = 1$, even though MCM surpasses SCM only in two of nine cases, it actually leads to the same or almost the same results in the remaining seven cases. This means that when segmentation is performed for every new ping when the corresponding swath vector is available, the way in which overlapping regions are fused is not particularly relevant, probably because there is so much information that the fusion method does not make the difference.

However, when it comes fo $\Delta t = \frac{N-1}{2}$, MCM surpasses SCM in all cases. The differences in this case are very small, but it is very significant that MCM is better independently on the training step pS and the class. Actually, the differences between this configuration and $\Delta t = N$ are larger for MCM than for SCM, showing how MCM is able to take profit of partially overlapping patches.

The F1-Scores are summarized in Figure 13. Similarly to the SCM case (Figure 12), results get worse and the differences between classes increase with the value of pS, thus encouraging the use of $pS = 1$. In this case, however, the effects of Δt are perfectly clear, since in all cases the F1-Score gets worse with Δt. This is due to the already mentioned improvement when $\Delta t = \frac{N-1}{2}$ using MCM with respect to the SCM case.

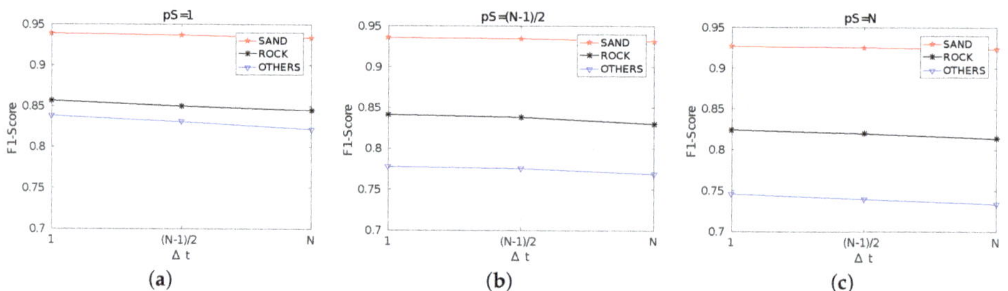

Figure 13. F1-Scores for MCM training with (**a**) $pS = 1$, (**b**) $pS = \frac{N-1}{2}$ and (**c**) $pS = N$.

Previous discussion about the effects of pS and Δt included the intuitive idea that small pS or Δt would increase computational requirements. In order to quantify this intuition, both the training and the segmentation times have been measured on the provided Python implementation, which relies on Keras using TensorFlow as backend, executed on a standard laptop endowed with an i7 CPU at 3.1 GHz and without using neither GPU nor TPU.

Table 8 shows the results, which are graphically summarized in Figure 14. The times, expressed in milliseconds, are the mean time per swath. More specifically, for each fold in the K-Fold cross-validation the training time has been measured and divided by the number of swaths or emitted pings in the training data corresponding to that fold. This training time has been averaged for all folds and the result is the training time per swath shown in the table.

Table 8. Training and segmentation time consumption when using SCM and MCM.

Time consumption (SCM)						
Δt / pS	1		$\frac{N-1}{2}$		N	
	Training	**Segment.**	**Training**	**Segment.**	**Training**	**Segment.**
1	1373.2110 ms	3.2084 ms	1373.2110 ms	0.2174 ms	1373.2110 ms	0.2148 ms
$\frac{N-1}{2}$	36.0540 ms	3.2304 ms	36.0540 ms	0.2295 ms	36.0540 ms	0.2294 ms
N	18.0790 ms	3.3044 ms	18.0790 ms	0.2323 ms	18.0790 ms	0.2145 ms
Time consumption (MCM)						
Δt / pS	1		$\frac{N-1}{2}$		N	
	Training	**Segment.**	**Training**	**Segment.**	**Training**	**Segment.**
1	1373.2110 ms	3.8411 ms	1373.2110 ms	0.6879 ms	1373.2110 ms	0.6580 ms
$\frac{N-1}{2}$	36.0540 ms	3.8038 ms	36.0540 ms	0.6799 ms	36.0540 ms	0.6163 ms
N	18.0790 ms	3.7451 ms	18.0790 ms	0.6960 ms	18.0790 ms	0.6328 ms

As for the segmentation time, a similar procedure has been used. In this case the measured time is not only the NN prediction time but also the times spent to build the patch to segment, to put the segmented patch into the segmented image and to compute the most probable class when necessary have also been measured.

Since training is not affected neither by the value of Δt nor by the use of SCM or MCM, the NN was trained only once per value of pS. That is why the training times are the same independently of Δt and the use of SCM or MCM, and that is the reason why a single plot of the training time as a function of pS is provided in Figure 14a. Results show how training time is particularly large when using $pS = 1$ and is drastically reduced by increasing the patch separation. However, since training has to be performed only once, it should not be an relevant criterion to select one configuration or another.

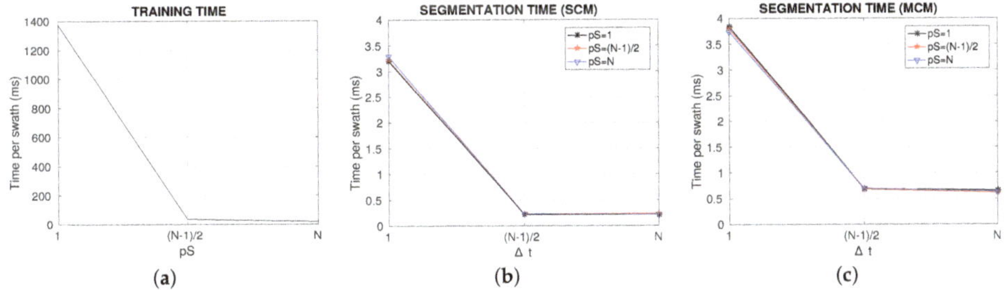

Figure 14. Execution times for (**a**) training, (**b**) segmenting using SCM and (**c**) segmenting using MCM.

Figure 14b,c clearly show that the segmentation time is not influenced by pS. This is reasonable, since pS only takes part in the training process. These figures also show a huge reduction in the segmentation time when switching from $\Delta t = 1$ to $\Delta t = \frac{N-1}{2}$ but an almost negligible reduction when going from $\Delta t = \frac{N-1}{2}$ to $\Delta t = N$. This is particularly interesting, since it means that choosing one of these two values of Δt can be done without taking the time into consideration.

By comparing SCM (Figure 14b) and MCM (Figure 14c) it is easy to see that, even though the segmentation times follow the same pattern, MCM is significantly more computationally demanding. For example, the smallest segmentation time when using SCM is 0.2148 ms whilst the smallest segmentation time with MCM is 0.6163 ms, which is almost three times larger.

Finally, it is important to emphasize that the segmentation times are really small in all cases. The worst situation, which happens when using MCM, $pS = 1$ and $\Delta t = 1$, requires 3.8411 ms per swath in average and the best one, which appears with SCM, $pS = N$ and $\Delta t = N$, uses 0.2145 ms per swath in average. This means that the system is able to process, depending on the configuration, between 260.342 and 4662.005 swaths per second in average. These frequencies are larger by far, in any case, to typical SSS sampling frequencies. For example, the used SSS provides 10 swaths per second, as shown in Table 1.

5.4. Qualitative Results

We conducted some experiments in order to visualize the effects of different Δt and the use of SCM or MCM to build the segmented acoustic image.

Figure 15a shows a fragment of a transect overlaid to the corresponding hand labelled ground truth. The black strip in the middle represents the blind zone, though, as explained before, it has not taken part in the segmentation process. The strips on the top and the bottom of the black region correspond to the port and starboard informative images respectively. Both informative images have been processed separately, and are shown here together to provide a clear representation of a segmented transect. The colors used to draw the ground truth are red to denote the rock class, blue to denote rippled sand and green to denote the class others.

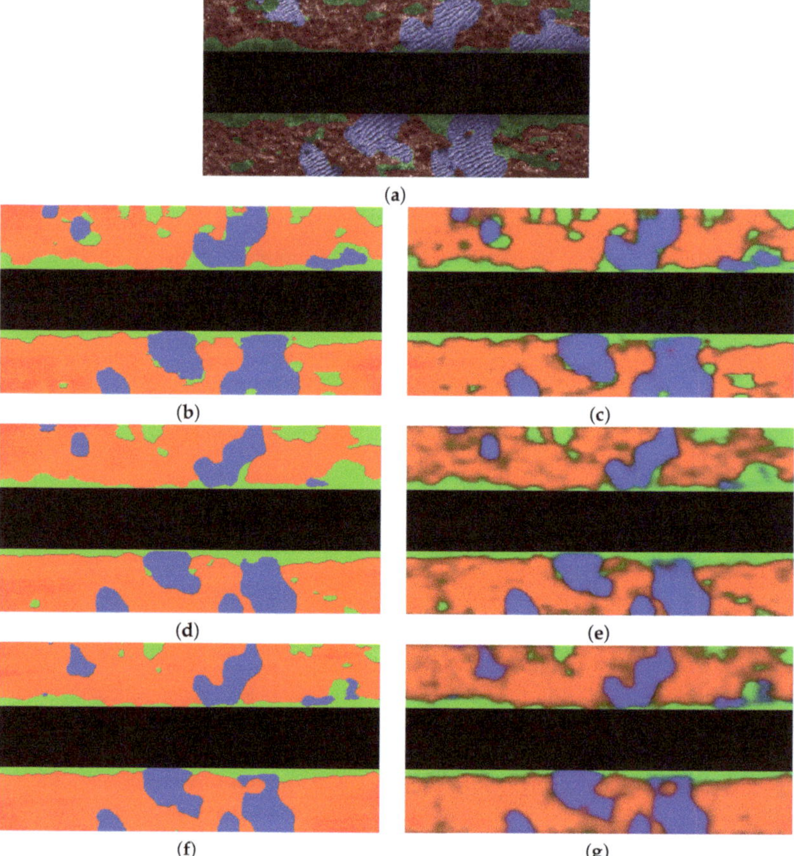

Figure 15. Segmentation results. (**a**) Informative images corresponding to a small transect overlaid with the ground truth and segmented images using (**b**) SCM and $\Delta t = 1$, (**c**) MCM and $\Delta t = 1$, (**d**) SCM and $\Delta t = 41$, (**e**) MCM and $\Delta t = 41$, (**f**) SCM and $\Delta t = 83$ and (**g**) MCM and $\Delta t = 83$.

Figure 15b–f show the resulting segmentation under different configurations after training our system with all the transects in the dataset except the one to which this example belongs. During training, the parameter pS ha been set to the intermediate value of $\frac{N-1}{2} = 41$. The effects of pS have already experimentally assessed in Section 5.3.

More specifically, Figure 15b–f show the results corresponding to the SCM using $\Delta t = 1$, $\Delta t = 41$ and $\Delta t = 83$ respectively. Being these results single-class, each bin is assigned a single label and, thus, the class boundaries are perfectly defined. The tested values of Δt range from performing a new segmentation every time a new swath is available ($\Delta t = 1$) to segmenting patches with no overlap at all ($\Delta t = 2 \cdot pM + 1 = 83$) as explained in Section 4.2.2. The qualitative effect of changing this parameter is a decrease in detail as Δt increases. For example, most of the small *others* regions (green) surrounding the *sand* regions (blue), which appear in the ground truth and are almost perfectly detected with $\Delta t = 1$ are not present when $\Delta t = 83$.

Figure 15c,e,g show the MCM results also using $\Delta t = 1$, $\Delta t = 41$ and $\Delta t = 83$. Results are similar to SCM except that a gradation between classes can be observed, especially in the contours of each region. Also, the small *others* regions mentioned before are now appreciable even when using non overlapping patches ($\Delta t = 83$).

There is a final remark to be done with respect to these qualitative results. Even though MCM using $\Delta t = 1$ seems to provide the best results, the time consumption has to be taken into account. Performing one segmentation every time a new swath is available may not be suitable depending on the computational capabilities of the on-board computer. As a matter of fact, the quantitative evaluation in Section 5.3 shows that the time consumption when using $\Delta t = 1$ is really large. Moreover, although MCM seems to be able to preserve some small details even with a large Δt, depending on the SLAM or mapping algorithm where this data has to be used, a single label per bin may be necessary and, thus, MCM may not be directly usable.

5.5. Discussion

Deciding the particular parametrization to use in real time operation has to take into account two factors that have been evaluated: the segmentation quality and the segmentation time. We believe that training time should not take part in the decision since training is performed only once. Also, it has been shown that the specific training has no effect on the segmentation time. Accordingly, since the best overall quality appears with $pS = 1$ our proposal is to use this particular patch separation during training.

In this case, the best quality appears with $\Delta t = 1$, both with SCM and MCM. However, this is also, with difference, the most computationally expensive case. Thus, if computational resources are limited, which is likely to happen in AUVs, larger values for Δt are advisable. Since no significant differences appear, neither in quality nor in time consumption, between $\Delta t = \frac{N-1}{2}$ and $\Delta t = N$ when using SCM, both options seem to be equally interesting. MCM is more computationally demanding, but it surpasses SCM for $\Delta t = \frac{N-1}{2}$. Actually, it leads to a quality similar to SCM with $\Delta t = 1$ with much lower computational cost.

Additionally, MCM has shown to provide more visual detail than SCM before selecting the most probable class. This means that it can generate maps which are more meaningful for human inspection and also that some localization and SLAM algorithms could take profit of that feature.

Overall, even though the final decision depends on the computational power of the on-board computer, an advisable parametrization seems to be $pS = 1$, $\Delta t = \frac{N-1}{2}$ and MCM. This means that, in our particular computer setup, an average of 0.6879 ms will be used to segment each swath, making it possible to process an average of 1453.7 swaths per second. Assuming a SSS similar to ours, which provides 10 swaths per second (see Table 1), this means that only a 0.69% of the CPU time will be spent segmenting the data, reaching an accuracy of 90.70% and F1-Scores of 0.85, 0.837 and 0.831 for classes rock, sand and others respectively.

Comparing our proposal to the results reported by other researchers is difficult, since the number of classes used as well as their meaning are different to ours and among them and also the provided quality measures are usually ad-hoc. However, the study by [35], which proposes a NN method, states accuracies ranging between the 58% and the 68%, thus accomplishing a hit ratio significantly below ours. Also, [29] reports the obtained confusion matrices showing that an 85.2% and a 93.53% of rock and floor bins, respectively, are properly detected. These two classes being similar to ours rock and other, it is safe to conclude that this NN behaves similarly to ours, being slightly inferior detecting rocks and slightly better when it comes to other.

However, contrarily to [29] and other the existing methods, our proposal has three additional advantages. On the one hand, our proposal is able to operate on-line, being responsible of an average CPU occupancy below 1%. On the other hand, our NN having less parameters it is trainable with less data, whereas the mentioned study has to deal with specific data augmentation techniques. Finally, our proposal is not only able to work with a relatively fast SSS, which operates at 10Hz whereas other approaches only deal with 1Hz SSS, but also tolerates low resolutions: our SSS only provides 500 bins, of which a significant part is discarded, whilst other approaches have only been tested with SSS which, at least, double the resolution of ours [28,29].

6. Conclusions and Future Work

In this paper we have presented a method to perform on-line segmentation of SSS data. The proposal performs three main steps. First, it pre-processes the data to take into account the particularities of SSS sensing. In this way, the main artifacts due to the uneven ensonification pattern and the sound attenuation with distance are reduced. Second, it uses a sliding window to group the most recently gathered swaths into overlapping patches. These patches are used to feed a CNN in charge of segmenting them. Third, it fuses the segmented patches into a consistent segmentation of the environment.

Thanks to that, each data bin provided by the SSS is assigned to one specific class. This segmentation has many applications, such as semantically mapping the environment, detecting archaeological or geological items or quantifying the presence of underwater algae or plants. Also, the segmented data is useful for a subsequent SLAM step, since having each bin classified into one specific class would make it easy to detect loops.

Several experiments have been conducted using real SSS data gathered in coastal areas of Mallorca (Spain). In these experiments different configurations of our proposal have been explored and quantitatively evaluated thus helping in the process of deciding the best setup. They show, for example, that accuracies larger than 90% can be achieved in a three-class scenario requiring less than the 7% of CPU on a standard laptop.

The documented source code as well as some datasets and trained models are publicly available at https://github.com/aburguera/NNSSS.

Future research pursues one main goal: to endow a SLAM system [25] with true loop detection capabilities with SSS data. Since data association using SSS is a difficult task due, among others, to significant changes in the received echoes depending on the viewpoint, including information about the classes of the existing sea floor parts will improve the ability to detect loops.

To this end, our proposal is to constrain the search for candidate loops to regions assigned to the same class. After that, a full and robust registration algorithm could be used to confirm or deny these loops. This would reduce the computational cost of data registration and, thus, help in achieving one of the main SLAM requirements, which is on-line operation.

Author Contributions: Both authors contributed equally to this work, including the Conceptualization, theoretical methodology, the implementation of the software, the validation with the datasets obtained in the sea, writing the original draft, and the final supervision. All authors have read and agreed to the published version of the manuscript.

Funding: This work is partially supported by Ministry of Economy and Competitiveness under contract DPI2017-86372-C3-3-R (AEI,FEDER,UE).

Acknowledgments: The underwater equipment used to gather this dataset was provided by Unidad de Tecnología Marina-CSIC (http://www.utm.csic.es/). The authors wish to thank Pablo Rodríguez Fornes, from UTM-CSIC, and Yvan Petillot, from Heriot-Watt University, for sharing their expertise with us and providing the data used in the experiments presented in this article. The authors are also grateful to Daniel Moreno Linares for his help with the XTF format.

Conflicts of Interest: The authors declare no conflict of interest.

References

1. Burguera, A.B.; Bonin-Font, F. A trajectory-based approach to multi-session underwater visual SLAM using global image signatures. *J. Mar. Sci. Eng.* **2019**, *7*. [CrossRef]
2. Köser, K.; Frese, U. Challenges in Underwater Visual Navigation and SLAM. In *Intelligent Systems, Control and Automation: Science and Engineering*; Springer: Cham, Switzerland, 2020; Volume 96, pp. 125–135. [CrossRef]
3. Wu, Y.; Ta, X.; Xiao, R.; Wei, Y.; An, D.; Li, D. Survey of underwater robot positioning navigation. *Appl. Ocean Res.* **2019**. [CrossRef]
4. Marage, J.P.; Mori, Y. *Sonar and Underwater Acoustics*; John Wiley & Sons Ltd.: Hoboken, NJ, USA, 2013. [CrossRef]
5. Le Bas, T.P.; Somers, M.L.; Campbell, J.M.; Beale, R. Swath bathymetry with GLORIA. *IEEE J. Ocean. Eng.* **1996**, *21*, 545–552. [CrossRef]
6. Searle, R.C.; Le Bas, T.P.; Mitchell, N.C.; Somers, M.L.; Parson, L.M.; Patriat, P. GLORIA image processing: The state of the art. *Mar. Geophys. Res.* **1990**, *12*, 21–39. [CrossRef]
7. Burguera, A. Underwater Localization using Probabilistic Sonar Registration and Pose Graph Optimization. In Proceedings of the 2018 IEEE/OES Autonomous Underwater Vehicle Workshop (AUV), Porto, Portugal, 6–9 November 2018. [CrossRef]
8. Marx, D.; Nelson, M.; Chang, E.; Gillespie, W.; Putney, A.; Warman, K. Introduction to synthetic aperture sonar. *IEEE Signal Process. Workshop Stat. Signal Array Process. SSAP* **2000**, 717–721. [CrossRef]
9. Ribas, D.; Ridao, P.; Neira, J. Understanding Mechanically Scanned Imaging Sonars. In *Underwater SLAM for Structured Environments Using an Imaging Sonar*; Springer: Berlin, Germany, 2010; pp. 37–46. [CrossRef]
10. Sousa-Sena, A.L. Shallow Water Remote Sensing Using Sonar Improved With Geostatistics and Stochastic Resonance Data Processing. Ph.D. Thesis, Universitat de les Illes Balears, Palma, Illes Balears, Spain, 2018.
11. Ji, D.; Liu, J. Multi-Beam Sonar Application on Autonomous Underwater Robot. *Mar. Geod.* **2015**, *38*, 281–288. [CrossRef]
12. Van Veen, B.D.; Buckley, K.M. Beamforming: A Versatile Approach to Spatial Filtering. *IEEE ASSP Mag.* **1988**, *5*, 4–24. [CrossRef]
13. Mallios, A.; Vidal, E.; Campos, R.; Carreras, M. Underwater caves sonar data set. *Int. J. Robot. Res.* **2017**, *36*, 1247–1251. [CrossRef]
14. Jiang, M.; Song, S.; Li, Y.; Jin, W.; Liu, J.; Feng, X. A Survey of Underwater Acoustic SLAM System. In *Lecture Notes in Computer Science*; Springer: Cham, Switzerland, 2019; Volume 11741 LNAI, pp. 159–170. [CrossRef]
15. Sternlicht, D.D. Historical development of side scan sonar. *J. Acoust. Soc. Am.* **2017**, *141*, 4041–4041. [CrossRef]
16. Savini, A. Side-Scan Sonar as a Tool for Seafloor Imagery: Examples from the Mediterranean Continental Margin. In *Sonar Systems*; IntechOpen Ltd.: London, UK, 2011. [CrossRef]
17. Burguera, A.; Oliver, G. High-resolution underwater mapping using Side-Scan Sonar. *PLoS ONE* **2016**, *11*. [CrossRef]
18. Johnson, H.P.; Helferty, M. The geological interpretation of Side-Scan Sonar. *Rev. Geophys.* **1990**. [CrossRef]
19. Bava-De-Camargo, P.F. The use of side scan sonar in Brazilian Underwater Archaeology. In Proceedings of the IEEE/OES Acoustics in Underwater Geosciences Symposium, Rio de Janeiro, Brazil, 29–31 July 2016. [CrossRef]
20. Cobra, D.T.; Oppenheim, A.V.; Jaffe, J.S. Geometric Distortions in Side-Scan Sonar Images: A Procedure for Their Estimation and Correction. *IEEE J. Ocean. Eng.* **1992**, *17*, 252–268. [CrossRef]

21. Sheffer, T.; Guterman, H. Geometrical Correction of Side-scan Sonar Images. In Proceedings of the 2018 IEEE International Conference on the Science of Electrical Engineering in Israel, ICSEE 2018, Eilat, Israel, 12–14 December 2018. [CrossRef]
22. Bikonis, K.; Moszynski, M.; Lubniewski, Z. Application of shape from shading technique for side scan sonar images. *Pol. Marit. Res.* **2013**, *20*, 39–44. [CrossRef]
23. Reed, S.; Petillot, Y.; Bell, J. Mine detection and classification in side scan sonar. *Sea Technol.* **2004**, *45*, 35–39.
24. Aulinas, J.; Lladó, X.; Salvi, J.; Petillot, Y.R. Feature based SLAM using side-scan salient objects. In Proceedings of the MTS/IEEE OCEANS, Seattle, WA, USA, 20–23 September 2010. [CrossRef]
25. Moreno, D.; Burguera, A.; Oliver, G. SSS-SLAM: An Object Oriented Matlab Framework for Underwater SLAM using Side Scan Sonar. In Proceedings of the XXXV Jornadas de Automática, Valencia, Spain, 3–5 September 2014.
26. Saini, K.; Dewal, M.L.; Rohit, M. Ultrasound Imaging and Image Segmentation in the area of Ultrasound: A Review. *Int. J. Adv. Sci. Technol.* **2010**, *24*, 41–60.
27. Priyadharsini, R.; Sharmila, T.S. Object Detection in Underwater Acoustic Images Using Edge Based Segmentation Method. *Procedia Comput. Sci.* **2019**, *165*, 759–765. [CrossRef]
28. Williams, D.P. Fast Unsupervised Seafloor Characterization in Sonar Imagery Using Lacunarity. *IEEE Trans. Geosci. Remote Sens.* **2015**, *53*, 6022–6034. [CrossRef]
29. Khidkikar, M.; Balasubramanian, R. Segmentation and classification of side-scan sonar data. In Proceedings of the Lecture Notes in Computer Science (Including Subseries Lecture Notes in Artificial Intelligence and Lecture Notes in Bioinformatics), Montreal, QC, Canada, 3–5 October 2012; Volume 7506 LNAI, pp. 367–376. [CrossRef]
30. Pinto, M.; Ferreira, B.; Matos, A.; Cruz, N. Side scan sonar image segmentation and feature extraction. In Proceedings of the MTS/IEEE Biloxi—Marine Technology for Our Future: Global and Local Challenges, OCEANS 2009, Biloxi, MS, USA, 26–29 October 2009.
31. Daniel, S.; Le Léannec, F.; Roux, C.; Solaiman, B.; Maillard, E.P. Side-scan sonar image matching. *IEEE J. Ocean. Eng.* **1998**, *23*, 245–259. [CrossRef]
32. Sultana, F.; Sufian, A.; Dutta, P. Evolution of Image Segmentation using Deep Convolutional Neural Network: A Survey. *Knowl.-Based Syst.* **2020**, *201–202*. [CrossRef]
33. Alhasoun, F.; Gonzalez, M. Streetify: Using Street View Imagery and Deep Learning for Urban Streets Development. In Proceedings of the IEEE International Conference on Big Data, Los Angeles, CA, USA, 9–12 December 2019; pp. 2001–2006. [CrossRef]
34. Van Opbroek, A.; Achterberg, H.C.; Vernooij, M.W.; De Bruijne, M. Transfer learning for image segmentation by combining image weighting and kernel learning. *IEEE Trans. Med. Imaging* **2019**, *38*, 213–224. [CrossRef]
35. Yu, F.; Zhu, Y.; Wang, Q.; Li, K.; Wu, M.; Li, G.; Yan, T.; He, B. Segmentation of Side Scan Sonar Images on AUV. In Proceedings of the 2019 IEEE Underwater Technology (UT), Kaohsiung, Taiwan, 16–19 April 2019; pp. 1–4. [CrossRef]
36. Coiras, E.; Petillot, Y.; Lane, D.M. Multiresolution 3-D reconstruction from side-scan sonar images. *IEEE Trans. Image Process.* **2007**, *16*, 382–390. [CrossRef] [PubMed]
37. Burguera, A. Segmentation of Side-Scan Sonar Data—Source Code. Available online: https://github.com/aburguera/NNSSS (accessed on 16 July 2020).
38. Burguera, A. A novel approach to register sonar data for underwater robot localization. In Proceedings of the Intelligent Systems Conference (IntelliSys 2017), London, UK, 7–8 September 2017; Volume 2018-January, pp. 1034–1043. [CrossRef]
39. Chang, Y.C.; Hsu, S.K.; Tsai, C.H. Sidescan sonar image processing: Correcting brightness variation and patching gaps. *J. Mar. Sci. Technol.* **2010**, *18*, 785–789.
40. Kleeman, L.; Kuc, R. Sonar Sensing. In *Springer Handbook of Robotics*; Springer: Berlin/Heidelberg, Germany, 2008; pp. 491–519. [CrossRef]
41. Greenspan, M.; Tschiegg, C.E. Tables of the Speed of Sound in Water. *J. Acoust. Soc. Am.* **1959**, *31*, 75–76. [CrossRef]
42. Abramowitz, M.; Stegun, I.A. *Handbook of Mathematical Functions: With Formulas, Graphs, and Mathematical Tables*; Applied Mathematics Series 55; U.S. Government Printing Office: Washington, DC, USA, 1964; pp. 591–592.

43. Langer, D.; Hebert, M. Building qualitative elevation maps from side scan sonar data for autonomous underwater navigation. In Proceedings of the 1991 IEEE International Conference on Robotics and Automation, Sacramento, CA, USA, 9–11 April 1991; Volume 3, pp. 2478–2483. [CrossRef]
44. Chollet, F. *Deep Learning with Phyton*; Manning Publications Co.: Shelter Island, NY, USA, 2018; p. 386.

Journal of
Marine Science and Engineering

Article

An Approach for Diver Passive Detection Based on the Established Model of Breathing Sound Emission

Qiang Tu [1], Fei Yuan [1], Weidi Yang [2] and En Cheng [1,*]

1 Key Laboratory of Underwater Acoustic Communication and Marine Information Technology, Ministry of Education, Xiamen University, Xiamen 361005, China; tuqiang@stu.xmu.edu.cn (Q.T.); yuanfei@xmu.edu.cn (F.Y.)
2 College of Ocean and Earth Sciences, Xiamen University, Xiamen 361102, China; wdyang@xmu.edu.cn
* Correspondence: chengen@xmu.edu.cn; Tel.: +86-139-5016-5480

Received: 12 December 2019; Accepted: 10 January 2020 ; Published: 15 January 2020

Abstract: Diver breathing sounds can be used as a characteristic for the passive detection of divers. This work introduces an approach for detecting the presence of a diver based on diver breathing sounds signals. An underwater channel model for passive diver detection is built to evaluate the impacts of acoustic energy transmission loss and ambient noise interference. The noise components of the observed signals are suppressed by spectral subtraction based on block-based threshold theory and smooth minimal statistic noise tracking theory. Then the envelope spectrum features of the denoised signal are extracted for diver detection. The performance of the proposed detection method is demonstrated through experimental analysis and numerical modeling.

Keywords: underwater acoustic signal processing; channel model; signal enhancement; signal denoising; passive detection

1. Introduction

A diver is an underwater swimmer who carries a self contained underwater breathing apparatus (SCUBA) system and can stay underwater for a long time. Because of the presence of water, people ashore find it difficult to find, to search for, and to communicate with divers. In addition, when a diver is in danger, the probability of misfortune is high, even with the help of rescuers. There are active and passive sonar system for underwater detection. In shallow water, the active sonar system faces the challenge of reverberation, and the performance requirements of small targets are high. Compared with the active mode, passive sonar has small energy consumption, is cheaper and more hidden, and is being pursued as an alternative [1].

In passive diver detection system, the diver's breathing sound, coming from the gas exchange process in SCUBA, is useful for the passive detection of the diver's presence [2,3]. The periodic pulse characteristic, caused by the vibration of high pressure gas in inhaling [4], is effective to detect the diver's presence. Ref. [5] proposed matched filter to extract periodic characteristic, but reliable reference signal from the diver's breathing sound is hard to obtain. Ref. [6] pre-whiten the noise and detect the diver based on envelope spectrum to a maximum range of 20 m. Although the sounds can be spatially filtered using an underwater array [7], we focus on detecting the presence of diver in a single channel, which also can be used in the multichannel scene.

The performance of passive detection is affected by the underwater environment, mainly including ambient noise interference and transmission loss. The noise spectrum in the ocean is colored by turbulence, rainfall, marine animals, and ships [8]. Since the diver-oriented sound spectrum distributes from hundreds of Hz to more than 75 kHz [7]. Diver detection is mainly affected by wind wave noise from the sea surface [9]. Another difficulty comes from the transmission loss, whose attenuation factors mainly include water absorption [5], geometric diffusion loss, bottom and surface scattering. In order to predict the characteristics of sound transmission, an acoustic rays model is mostly adopted [10].

Due to low signal to noise ratio (SNR) of observed signals, noise suppression is necessary for detection system, includes noise spectral estimation and noise removing steps. There are many ways used to estimate noise spectral power. Minimum statistics algorithm tracks the minima values of a smoothed power estimate of the noisy signal [11]. Cohen further combined the minimum tracking and the recursive averaging, proposed minima-controlled recursive averaging algorithm (MCRA) [12] and improved algorithm (IMCRA) [13]. Hendriks proposed the subspace noise tracking algorithm (SNT) [14] to search for the signal dimension number and to estimate the noise spectral power in each subspace. Then, the IMCRA method is adopted because of good performance under low SNR conditions [15]. To remove noise from noisy signals, the block-based threshold algorithm (BT) [16] is adopted. Compared with others noise suppression methods, such as random matched filtering [17], cepstral minimum mean-square error motivated noise suppress [18], wavelet threshold [19], the BT method can adaptively estimate the best noise reduction coefficient on time-frequency point at low SNR [20]. The BT method minimizes Stein's unbiased risk estimator (SURE) [21,22] to obtain adaptive block area size and threshold level. It means that the estimated attenuation coefficients of center point in blocks are the results of operation of others points in the blocks.

The present work will focus on diver passive detection, and underwater acoustic channel model from sound source to hydrophone. Firstly, the model of transmission loss and ambient noise is built to evaluate the measured SNR of observed diver's breathing sounds. Secondly, we introduce an adaptive noise subtraction approach to enhance the diver's breathing sounds, which does not need prior knowledge of signals. The ambient noise is suppressed by spectral subtraction approach which is based on BT theory and IMCRA method. Then, extract the envelope spectrum of diver breathing signal for basis feature of diver detection. Finally, detection performance is proved by practical experiment and numeral analysis.

The rest of the paper is organized as follows. Section 2 introduces the acoustic channel model about transmission loss and ambient noise. Section 3 presents detection approach algorithm including noise estimation algorithm, BT algorithm for noise subtraction, envelope spectrum detection method. In Section 4, data acquisition experiment and source signal analysis are introduced. Then, Section 5 evaluates the SNR of measurement of diver signals through underwater channel and the performance of the noise subtraction for detection. Finally, the conclusions are given in Section 6.

2. Underwater Acoustic Channel Model

In underwater acoustic environments, the relationship between received sound level (RL) and source sound level (SL) follows passive sonar equation $RL = SL - TL + NL$. SL represents the diver breathing sound level, is related to measuring in standard range (1 m). TL is transmission loss and NL is ambient noise level at hydrophone. As Figure 1 shows, transmission loss and ambient noise are the main parts of underwater acoustic channel model for diver detection.

The acoustic energy transmission loss of the diver breathing soundwave is divided into three kinds as geometric diffusion loss, water absorption loss and scattering loss. In order to predict the transmission loss, the normal mode model and the ray model are often used to model the acoustic transmission process. Considering that the ray model is more suitable for simulating the scene of high frequency signal detection in short distance, we use it to model the underwater transmission of diver breathing sounds. The received signal $R(t)$ can be expressed as

$$R(t) = \sum_{i=1}^{L} \alpha_i A_i \delta(t - \tau_i) \tag{1}$$

where L is the number of intrinsic rays, A_i is the amplitude of ith ray and α_i represents attenuation coefficient. τ_i is the time delay of each ray. Diver breathing sound is regarded as a point sound source, and the sound wave diffuses in the form of spherical wave, that is, geometric diffusion loss. Water absorption loss is related to the temperature, salinity, PH, frequency, the distance of hydrophone. An experience formula Thorp [5] of predicting the absorption coefficient can be expressed as

$$\alpha(f) = \frac{0.1f^2}{1+f^2} + \frac{40f^2}{4100+f^2} + 2.75 \times 10^{-4} f^2 + 0.003 \tag{2}$$

where f is signal frequency in kHz. Scattering attenuation is due to the scattering of sound waves by the uneven and rough surface of the sea bottom and the sea surface, which leads to the attenuation of sound waves.

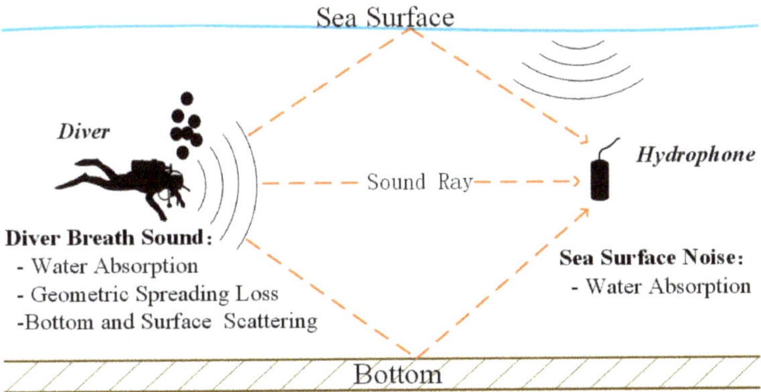

Figure 1. Underwater acoustic channel model for diver detection. Transmission loss contains geometry diffusion loss, water absorption and scattering by bottom and surface. Observed signals are affected by ambient noise, for example, wind noise from sea surface.

Besides, ambient noise is also essential in underwater acoustic channel model. Wind noise and ship noise are the main noise in ambient noise. The frequency of the diver's breathing sound we are concerned about is more than 2 kHz. While the ship noise spectrum power is mainly distributing below 200 Hz [23], the ship noise can be ignored. The ambient noise is mainly wind noise above 1 kHz [24]. The wind noise is caused by the vibration of bubbles when the waves hit the sea surface. The designed noise generator uses logarithmic relationship between wind speed and ambient noise level, which is given as [25]

$$\log N_w(f) = 5 + 0.75w^{1/2} + 2\log f - 4\log(f+0.4) \tag{3}$$

where f denotes sound frequency in Hz, w is wind speed in m/s, N_w is ambient noise level in dB. In the process of transmission, wind noise is also affected by water absorption attenuation. If the scattering of sound waves from the bottom of water is ignored, the transmission loss of wind noise is expressed as [26]

$$TL_{noise} = \alpha_w \times d \tag{4}$$

where TL_{noise} denotes the transmission loss of wind noise in dB, α_w is the attenuation coefficient in dB/km, d is the hydrophone depth in km.

3. Noise Reduction and Detection Methodology

This section describes the diver detection process, including noise suppression theory and envelope spectrum detection theory. The framework of proposed diver detection method demonstrates in Figure 2.

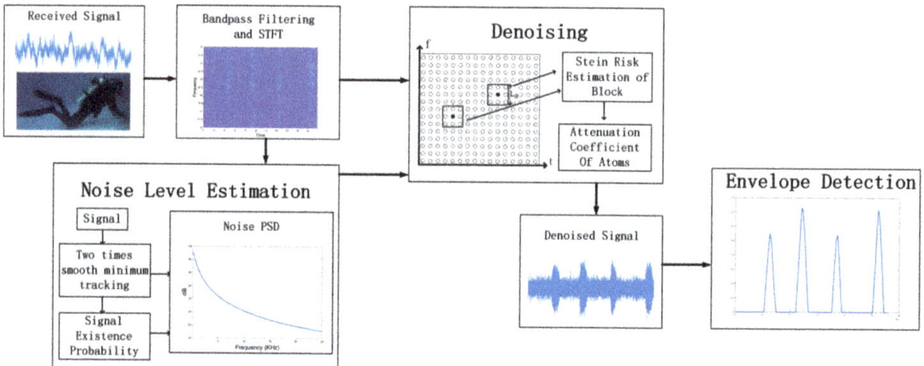

Figure 2. Framework of diver detection method.

3.1. Noise Reduction

Set y as observed time series of noisy signals. By short time Fourier transform (STFT), time series are decomposed into a family of time-frequency atoms $Y(k,l)$, where k and l are time and frequency scale. In time-frequency domain, the principle of spectral subtraction is to shrink time-frequency points by attenuation coefficient α_{kl}. The purpose of α value design is to remove the noise components and keep the signal components. Then, the enhanced signal in time-frequency domain $\widetilde{Y}_{kl}$ is given as

$$\widetilde{Y}_{kl} = \alpha_{kl} Y_{kl} \tag{5}$$

To obtain effective α_{kl}, surrounding points of $Y(k,l)$ are divided into a block area. Then, the α_{kl} is given as

$$\alpha_{kl} = (1 - \frac{\lambda}{\gamma_{B_{kl}}})_+ \tag{6}$$

where $\lambda > 0$ denotes the threshold that decides signals presence or not, operation $(g)_+ = max(g,0)$, B_{kl} is block area at point (k,l). Assuming noise power is known and is δ^2, γ is the posterior SNR which is given as $\gamma_{kl} = Y^2(k,l)/\delta^2$. Equation (6) demonstrates that the denoising performance of the α is related to block size L_B and threshold level λ. Because pure reference signal Y_{pure} is unknown, the Stein unbiased risk estimation (SURE) [21] algorithm is used to estimate risk equation given as [16]

$$\begin{aligned} \widetilde{R}_i &= \sum_{l,k \in B_i} E|Y_{pure}[k,l] - a_i Y[k,l]|^2 \\ &\underline{\underline{SURE}}\ L_B^2 + \sum_{n=1}^{L_B^2} ||h_n(\gamma_n)||^2 + 2\sum_{n=1}^{L_B^2} \frac{\partial h_n(\gamma_n)}{\partial \gamma_n} \end{aligned} \tag{7}$$

where γ_n denotes nth point in block B_i. Function $h_n(\gamma_n)$ is given as

$$h_n(Y_n) = S_n - Y_n = \begin{cases} -\frac{\lambda^2}{S_n^2} \cdot \gamma_n & (S_n > \lambda) \\ -\gamma_n & (S_n \le \lambda) \end{cases} \tag{8}$$

where $S_n = \alpha_n Y_n$. Then, the square equation and the derivative equation of h_n are given as

$$|h_n\,(Y_n)\,|_2^2 = \begin{cases} \frac{\lambda^4}{S_n^4}(\frac{Y_n}{\sigma_n}))^2 & (S_n > \lambda) \\ (\frac{Y_n}{\sigma_n})^2 & (S_n \le \lambda) \end{cases} \tag{9}$$

$$\frac{\partial h_n(\gamma_n)}{\partial \gamma_n} = \begin{cases} -\lambda^2 \frac{S_n^2 - 2\gamma_n^2}{S_n^4} \cdot \gamma_n^2 & (S_n > \lambda) \\ -1 & (S_n \leq \lambda) \end{cases} \tag{10}$$

In Equation (7), the SURE risk is close to the minimum value in the iterative of B_i. The block size L_B must be close in the way that the signal and the noise have slow variations inside the blocks. If the noise is color, e.g., ocean ambient noise, the risk estimator can be near unbiased with a narrow frequency band block [16].

3.2. Noise Level Estimation

The discussion in the previous section assumed the noise level to be known. However, the prior information of ambient noise can not be known. We use the IMCRA approach [13] to get the posterior estimation of noise level. In time-frequency domain, the noise power σ^2 is estimated from statistical average of the noise spectrum power of the past time scale, which is given as

$$\widetilde{\sigma}_d^2(k, l+1) = \widetilde{\alpha}_d(k,l)\widetilde{\sigma}_d^2(k,l) + (1 - \widetilde{\alpha}_d(k,l))|Y(k,l)|^2 \tag{11}$$

where $\hat{\alpha}_d(k,l)$ denotes time-varying and frequency independent smooth parameter, which is given as

$$\widetilde{\alpha}_d(k,l) = \alpha_d + (1 - \alpha_d)p(k,l) \tag{12}$$

where α_d denotes scalar smoothing parameter, $p(k,l)$ is the presence probability of useful signals, which is given as

$$p(k,l) = (1 + \frac{q(k,l)}{1 - q(k,l)}(1 + \xi(k,l))\exp(-v(k,l)))^{-1} \tag{13}$$

where $q(k,l)$ denotes signal absence probability, $v(k,l) = frac\gamma\xi 1 + \xi$, γ and ξ are the posterior SNR and priori SNR, which are given as

$$\gamma(k,l) = \frac{|Y(k,l)|^2}{\sigma_d^2(k,l)} \tag{14}$$

$$\xi(k,l) = \alpha G_{H_1}^2(k, l-1)\gamma(k, l-1) + (1 - \alpha)\max\{\gamma(k,l), 0\} \tag{15}$$

where α denotes a weighting factor controlling the balance between noise reduction and signal distortion, G_{H_1} is spectral gain function. To estimate $p(k,l)$ robust, signal absence probability $q(k,l)$ is estimated by two iterations of smoothing and minimum tracking. The smoothing in iterations takes into account the strong correlation of neighboring frames in independent frequency bins by a first-order recursive averaging. In first iteration, frequency smoothing of each frame is defined by

$$S(k,l) = \alpha_s S(k, l-1) + (1 - \alpha_s)S_f(k,l) \tag{16}$$

where $\alpha_s(0 < \alpha_s < 1)$ denotes smoothing parameter for adjacent frame, $S_f(k,l)$ is the spectrum power of the noisy signal given as

$$S_f(k,l) = \sum_{i=-w}^{w} b(i)|Y(k-i,l)|^2 \tag{17}$$

where b is a normalized window function of length $2w + 1$, e.g., Hanmming window. Then, track the local minimal frequency bins in consecutive time frame with a window size D, which is given as

$$S_{min}(k,l) = minS(k,l')|l - D + 1 <= l' <= l \tag{18}$$

In the first iteration, a rough estimation of signal presence $I(k,l)$ is defined as

$$\mathrm{I}(k,l) = \begin{cases} 1, \ if \ \gamma_{min}(k,l) < \gamma_0 \text{ and } \zeta(k,\ l) < \zeta_0, (signal\ is\ absent) \\ 0, \qquad \text{otherwise}(signal\ is\ present) \end{cases} \tag{19}$$

where γ_0 and ζ_0 is threshold that use $\gamma_0 = 4.6$ and $\zeta_0 = 1.67$ typically. γ_{min} and ζ denote posterior SNR and priori SNR in minima tracking of first iteration, which are given as

$$\gamma_{\min}(k,\ \ell) = \frac{|Y(k,\ \ell)|^2}{B_{\min}S_{\min}(k,\ \ell)};\quad \zeta(k,\ \ell) = \frac{S(k,\ \ell)}{B_{\min}S_{\min}(k,\ \ell)}. \tag{20}$$

where B_{min} is the bias of minimum estimation. Then, in the second iteration, the smoothing process is similar with the first iteration. The spectrum power of the noisy signal is installed as

$$\tilde{S}_f(k,l) = \begin{cases} \frac{\sum_{i=-w}^{w} b(i) I(k-i,l)|Y(k-i,l)|^2}{\sum_{-w}^{w} b(i) I(k-i,l)} \\ \tilde{S}(k,l-1), otherwise \end{cases} \tag{21}$$

The signal absence probability $\widetilde{q}(k,l)$ is equation of updated γ_{min} and ζ, as

$$\hat{\mathrm{q}}(k,l) = \begin{cases} 1 \quad , \ if \ \tilde{\gamma}_{min}(k,l) \le 1 \ and \ \tilde{\zeta}(k,l) < \zeta_0 \\ \frac{(\gamma_1 - \tilde{\gamma}_{min}(k,l))}{(\gamma_1 - 1)}, \ if \ 1 < \tilde{\gamma}_{min}(k,l) < \gamma_1 \ and \ \tilde{\zeta}(k,l) < \zeta_0 \\ 0 \quad , \qquad \text{otherwise} \end{cases} \tag{22}$$

where $\tilde{\gamma}_{min}$ and $\tilde{\zeta}$ denote posterior SNR and priori SNR in minima tracking of second iteration. γ_1 is threshold that use $\gamma_1 = 3$ typically. In Equation (22), the threshold processing of $\tilde{\gamma}_{min}$ and $\tilde{\zeta}$ guarantees the performance of ambient noise estimation in the presence of weak signals.

3.3. Detection Method

Previous research has shown that frequency sub-band envelope spectrum detection (ESD) is an effective detection method to detect the presence of diver [3,6]. ESD takes D_{env} as the feature of the diver's breathing sound, where D_{env} denotes envelope spectrum energy in the range of typical human breathing rates 0.3 Hz–1 Hz. D_{env} takes large value when diver is present, otherwise takes small value. Because ambient noise not affect the envelope spectrum in the range of 0.3 Hz–1 Hz, D_{env} is useful even in the severe ambient noise [3].

Figure 3 shows the calculation process of D_{env}. We first extract the envelope of noise-reduced signal. The envelope has obvious periodic characteristic if diver can be detected, otherwise the envelope is random and irregular. Secondly, we transform the envelope into a spectrum. The periodic characteristic of the envelope has a related peak in the spectrum. Since human breathing rates vary with the human body state, e.g., fast swimming or slow swimming, the peak can appear in each position of typical human breathing rates 0.3 Hz–1 Hz. Then, integrate spectrum over 0.3 Hz–1 Hz range to calculate D_{env} for detection.

The results of detection are represented by detection probability P_D, which is given as

$$P_D = \begin{cases} 1, & if \quad D_{env} > 2T \\ \frac{D_{env} - T}{D_{env}}, & if \quad T < D_{env} <= 2T \\ 0, & if \quad D_{env} <= T \end{cases} \tag{23}$$

where T denotes threshold of diver detection. The selection of detection threshold is related to the level of ambient noise. We use the $T = D_{env}^N + \varepsilon$, where D_{env}^N is calculated by the noise signal, ε denotes a positive constant.

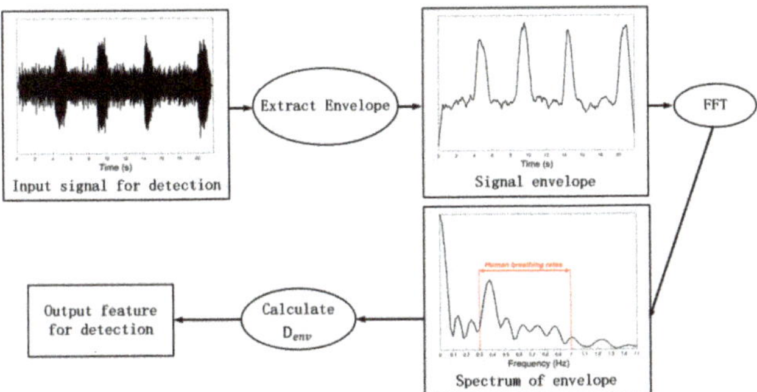

Figure 3. Flow chart of calculating D_{env} from signals.

Algorithm 1 Diver detection algorithm BIED based on BT and IMCRA.

Input: Observed signal
STEP 1: Bandpass filtered signal and STFT. Separate signal into many time frames.
STEP 2: For time frame l, compute posterior SNR $\gamma(k,l)$ as Equation (14) and prior SNR $\tilde{\xi}(k,l)$ as Equation (15)
STEP 3: Compute the first iteration of smoothing power spectrum $S(k,l)$ as Equations (16) and (17), track the minimum $S_{min}(k,l)$ as Equation (18).
STEP 4: Compute minima tracking noise's posterior SNR γ_{min} and priori SNR ξ as Equation (20).
STEP 5:Compute a roughly decision about signal presence $I(k,l)$ as Equation (19).
STEP 6: Install noise power spectrum $\widetilde{S}_f(k,l)$ as Equation (21).
STEP 7: Repeat the STEP 3–4.
STEP 8: Compute signal absence probability $\tilde{q}(k,l)$ as Equation (22). Compute signal presence probability $p(k,l)$ as Equation (13).
STEP 9: Compute smooth parameters $\tilde{\alpha_d}(k,l)$ as Equation (12).
STEP 10: Estimate noise power σ^2 as Equation (11).
STEP 11: Compute $h_n(\gamma n)$, $|h_n(\gamma_n)|_2^{(}2)$, $\partial h_n/\partial(\frac{\gamma_n}{\sigma_n})$ as Equations (8)–(10).
STEP 12: Compute risk in ith block as Equation (7), estimate threshold λ and block size L_B by iteration in blocks.
STEP 13: Compute attenuation coefficient $\alpha_{k,l}$ of atoms in time-frequency plane as Equation (6), obtain denoising signal $\widetilde{Y}_{kl}$ as Equation (5).
STEP 14: Transform the time-frequency representation into time series by inverse STFT.
STEP 15: Extract the envelope form result signals. Calculate D_{env} on envelope spectrum from 0.3 Hz to 1 Hz.
STEP 16: Calculate detection probability using Equation (23).
Output: Probability of the diver's presence.

In summary, the proposed diver detection method reduces noise based on BT and IMCRA, detecting the diver by feature from an envelope spectrum. We call it the BIED method. The detailed steps of the detection algorithm is shown in Algorithm 1.

4. Data and Analysis

The data of diver breathing sounds is collected in the swimming pool. The diver assisting in the experiment has more than five years of diving experience. In the experiment, a data acquisition card and a hydrophone were used to record underwater sounds. Figure 4 shows the diver equipped with

SCUBA system breaths underwater. The hydrophone is about 1m away from the diver. The sample rate is 50 kHz.

Figure 4. In experiment, one channel data acquisition system is used to record the diver's breathing sound underwater. Sample rate is 50 kHz.

Diver breathing sounds come from the air flow in the SCUBA system. The air flow process is controlled by the diver breath. The time series of the diver's breathing sound clearly shows the whole breathing process as Figure 5a shows. Through 2 kHz high pass filter and low pass filter, the inhaling and exhaling sounds can be separated as Figure 5b,c show. In Figure 6, the inhaling sounds frequency distribute in the range of 2 kHz–25 kHz. The frequency of exhaling sounds is mainly below 2 kHz. The inhaling sound and the exhaling sound can represent the diver's breathing process separately. Since the inhaling sounds have better pulse characteristic, while the waveform of exhaling sound is irregular. We use inhaling sound as the interested signal to diver detection.

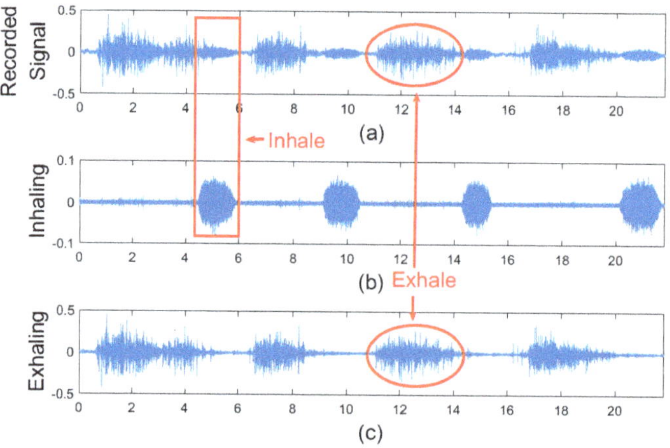

Figure 5. Breathing Sound recorded in experiment. The inhaling and exhaling sound are separated by high-pass and low-pass filters with 2 kHz cutoff frequency. (**a**) original recorded signal; (**b**) high frequency inhaling part of signal; (**c**) low frequency exhaling part of signal.

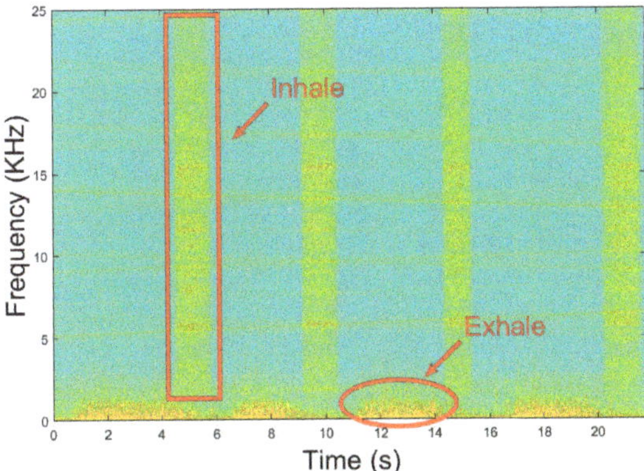

Figure 6. The spectrum of the diver's breathing sound. Inhaling sound frequency distributes in 2 kHz–25 kHz when sample rate is 50 kHz and exhaling sound frequency power is below 2 kHz.

5. Results and Analysis

5.1. Impacts of Underwater Environment

The main impacts of the underwater environment on diver detection are transmission loss and ambient noise interference. The above impacts are taken into account in the established underwater acoustic channel model for diver detection. Then, we can observe the change of breathing sound with channel parameters. Because the diver breathing sounds collected in the experiment have very obvious human breath rate characteristics, we regard them as source signals. Transmission loss is considered to be the result of geometry diffusion loss and water absorption loss. Because scattering attenuation has little effect on signal strength in short distance, we ignored scattering loss caused by bottom and surface. The diver detection environment is set as follows, source depth and receiver depth are 5 m, seafloor depth is 100 m, ambient noise related wind speed is 5 m/s. The Bellhop tool [27] is applied to calculate the attenuation coefficient of independent frequency. In the operations of Bellhop, the sound is modeled as Gaussian rays and is tracked by the sound rays at different incident angles from -80° to 80°. The ambient noise is considered to be slowly changing, and the associated sea surface wind speed is 5 m/s.

In Figure 7, the power spectral density (PSD) of source sound and attenuated sounds at the distance of 10 m, 30 m, 100 m are shown. With the increase of distance, the sound intensity of diver breathing sound decreases fast. At a distance of 100 m, the attenuation coefficient is close to 35 dB. Compared with the source signal, the acoustic signal attenuates nearly 20 dB at the distance of 10 m, nearly 30 dB at the distance of 30 m. That means the trend of sound intensity attenuation decreases exponentially. Therefore, transmission loss is mainly due to geometry diffusion loss in 100 m, and frequency dependent water absorption loss has little effect on signal attenuation. The frequency is not a major limitation in selecting sub-band for diver detection in 100 m.

Figure 8 shows the ambient noise, source sound and observed signals at the distance of 10 m, 30 m, and 100 m. Because of the effect of strong noise and strong attenuation, the observed signals have lost the waveform of source sound even at the distance of 10 m. Therefore, the first task of detection is to find the significant sub-band of the signal. The observed signals are divided into several sub-bands to discuss the effects of attenuation and noise, including 3 kHz–8 kHz, 8 kHz–13 kHz, 13 kHz–18 kHz and 18 kHz–23 kHz. Figure 9 compares the SNR of each sub-band. The SNR of sub-band 3 kHz–8 kHz is the lowest because the PSD of ambient noise is high in this frequency band. Otherwise, the SNR of other sub-bands are similar. We choose sub-band 13 kHz–18 kHz for diver detection because of the higher SNR.

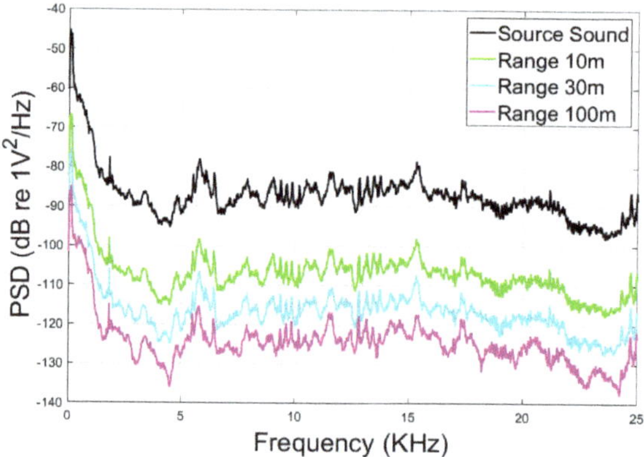

Figure 7. PSD of source sound and observed signals at the range of 10 m, 30 m, 100 m.

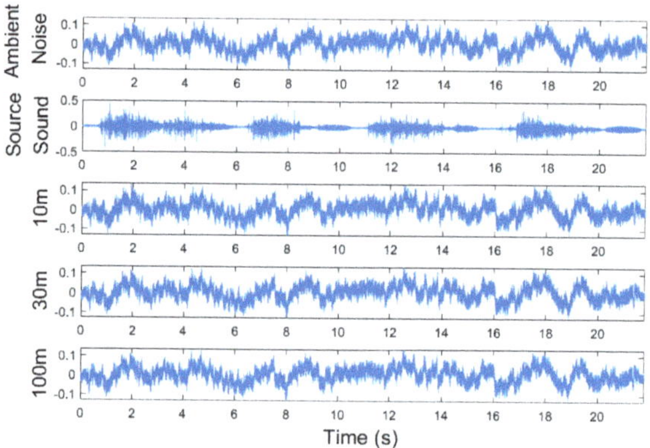

Figure 8. Ambient noise, source sound and observed signals at the range of 10 m, 30 m, 50 m.

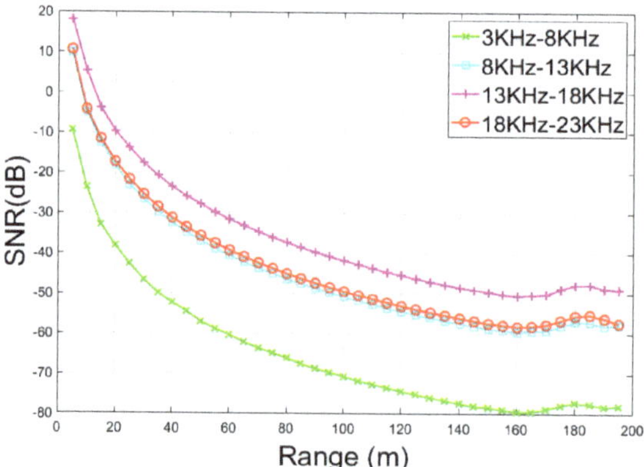

Figure 9. SNR of frequency band 3 kHz–8 kHz, 8 kHz–13 kHz, 13 kHz–18 kHz and 18 kHz–23 kHz. The 13 kHz–18 kHz band has the best SNR performance.

5.2. Detection System Performance

The detection of the underwater diver is affected by the underwater environment. For example, in a river or harbor, the environmental noise will cause the received SNR to decrease. We verify the performance of the detection system by adjusting the SNR. It is assumed that the ambient noise level is controlled by the wind and waves noise with 5 m/s wind speed, and the SNR can be changed by changing the detection distance. The proposed BIED method firstly uses SME theory and BT theory to estimate the ambient noise level and to remove the noise. Then, extract the characteristic value D_{env} from the envelope spectrum to detect the presence of a diver. The threshold of diver detection is set to $T = D_{env}^{N} + D_{env}^{N}/3$.

To evaluate the SNR of the denoised signal, an evaluation value SNR_M is defined as

$$SNR_M = 10 \log \frac{\Sigma y(n) \times M(n)}{\Sigma y(n) \times |M(n) - 1|} \tag{24}$$

where M denotes the manually marked presence position of diver breathing sounds, $|M - 1|$ is the opposite of M. In sequence M, the signal presence position is marked as 1, otherwise 0. The SNR_M represents the ratio of diver breathing sound presence signal component and absence signal component in time series. High SNR means that the envelope characteristics of diver breathing sound are more obvious and the D_{env} is high.

The length of time series also affects D_{env}. Theoretically, the larger the number of diver's breathing cycles contained in the observation window, the larger the corresponding detection value D_{env}. However, the long observation window does not meet the real detection requirement with reliability and timeliness. For example, when a diver is escaping from the hydrophone, a short window must be used to capture the presence of the diver in time. Hence, we use a time window of 22 s to detect diver, which contains four breathing periodic pulse at least.

Figure 10 compares the pre-processed signals of ESD method and the ones of proposed BIED method at the distance of 10 m and 30 m. The pre-processed signal of BIED has stronger inhaling sound pulse than the ESD's in high SNR condition as Figure 10a,b show. At the distance of 30 m, Figure 10d shows that the enhanced signal in BIED has inhaling sound characteristics, while the observed signal in ESD is almost submerged by noise as Figure 10c shows.

In Figure 11, the SNR_M of pre-processed signals in the ESD method and the proposed BIED method are compared. The curve of BIED method has higher SNR_M value than the curve of ESD method within a distance of less than 55 m. That proves the noise elimination process in BIED is effective to enhance the observed diver breathing sound. In the low SNR conditions, the noise elimination method is difficult to distinguish the background noise component form the observed signals. Then, two methods have approximate SNR_M value at a long distance.

In Figure 12, two curves show that the detection probability decreases as detection distance increases. The proposed BIED method has a higher detection probability in the near range. The reason for this is that the noise reduction process further enhances the SNR of 13 kHz–18 kHz band signal. The ESD method detects the diver to a maximum range of near 20 m, which is similar to the detection results of Johansson [6]. Compared with that, the BIED method can detect diver until the 40 m range.

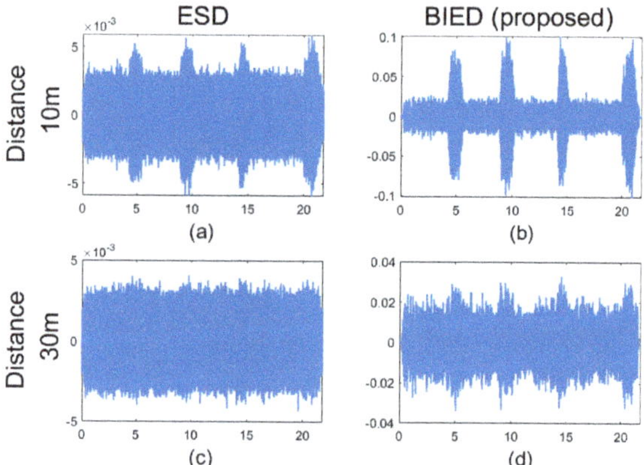

Figure 10. Pre-processed signals in the ESD method and the BIED method. (**a**) ESD at the distance of 10 m; (**b**) BIED at the distance of 10 m; (**c**) ESD at the distance of 30 m; (**d**) BIED at the distance of 30 m.

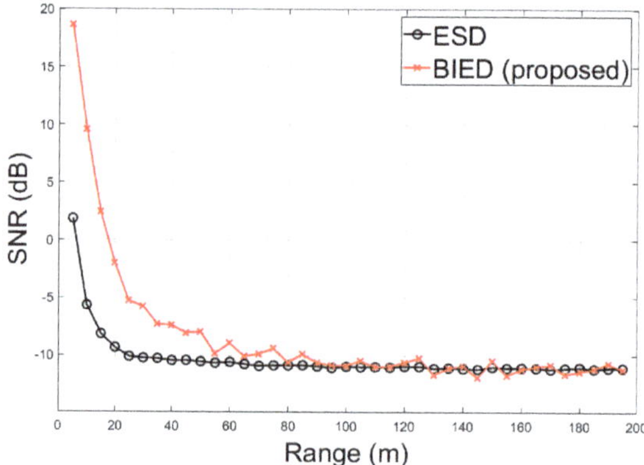

Figure 11. SNR of pre-processed signals in the ESD method and the BIED method.

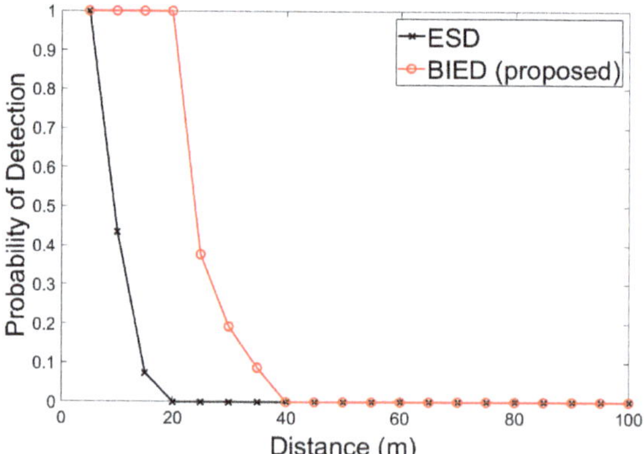

Figure 12. Detection probability. The detection threshold is set to $T = D_{env}^N + D_{env}^N/3$.

6. Conclusions

In this paper, we propose a diver detection method BIED based on suppressing ambient noise and extracting envelope spectrum features. The built acoustic channel model mainly considers transmission loss and noise interference in the underwater passive detection scenario. In the numeral analysis, the 13 kHz–18 kHz band of observed signals is selected for diver detection. While the ESD method can detect a range up to 20 m, the proposed BIED method detects one diver to a maximum range near 40 m.

Although our work shows effectiveness in diver detection, there are still many challenges to face. One of them is that the strength of the target sound source is too weak and easily covered by noise, which is the mainly reason for limiting detection distance. There is also a need to detect multiple divers' presences. We are working to achieve passive detection in these challenges.

Author Contributions: Conceptualization, Q.T. and F.Y.; methodology, Q.T.; software, Q.T.; validation, Q.T.; formal analysis, Q.T.; investigation, Q.T.; resources, F.Y. and Q.T.; data curation, W.Y. and F.Y. and Q.T.; writing–original draft preparation, Q.T.; writing–review and editing, Q.T.; visualization, Q.T.; supervision, F.Y.; project administration, E.C.; funding acquisition, E.C. All authors have read and agreed to the published version of the manuscript.

Funding: This research was funded by National Natural Science Foundation of China (61571377, 61771412, 61871336) and the Foundamental Research Funds for the Central Universities (20720180068).

Conflicts of Interest: The authors declare no conflict of interest.

References

1. Lennartsson, R.; Dalberg, E.; Johansson, A.; Persson, L.; Petrović, S.; Rabe, E. Fused passive acoustic and electric detection of divers. In Proceedings of the IEEE 2010 International WaterSide Security Conference, Carrara, Italy, 3–5 November 2010; pp. 1–8.
2. Stolkin, R.; Sutin, A.; Radhakrishnan, S.; Bruno, M.; Fullerton, B.; Ekimov, A.; Raftery, M. Feature based passive acoustic detection of underwater threats. In *Photonics for Port and Harbor Security II*; International Society for Optics and Photonics, Defense and Security Symposium: Orlando, FL, USA, 2006; Volume 6204, p. 620408.
3. Stolkin, R.; Florescu, I. Probabilistic analysis of a passive acoustic diver detection system for optimal sensor placement and extensions to localization and tracking. In Proceedings of the OCEANS 2007 MTS/IEEE, Vancouver, BC, Canada, 29 September–4 October 2007; pp. 1–6.
4. Donskoy, D.M.; Sedunov, N.A.; Sedunov, A.N.; Tsionskiy, M.A. Variability of SCUBA diver's acoustic emission. In *Optics and Photonics in Global Homeland Security IV*; International Society for Optics and Photonics, Defense and Security Symposium: Orlando, FL, USA, 2008; Volume 6945, p. 694515.
5. Harris, A.F., III; Zorzi, M. Modeling the underwater acoustic channel in ns2. In Proceedings of the 2nd International Conference on Performance Evaluation Methodologies and Tools, Nantes, France, 22–27 October 2007; p. 18.
6. Johansson, A.; Lennartsson, R.; Nolander, E.; Petrovic, S. Improved passive acoustic detection of divers in harbor environments using pre-whitening. In Proceedings of the OCEANS 2010 MTS/IEEE, Seattle, WA, USA, 20–23 September 2010; pp. 1–6.
7. Hari, V.N.; Chitre, M.; Too, Y.M.; Pallayil, V. Robust passive diver detection in shallow ocean. In Proceedings of the OCEANS 2015 MTS/IEEE, Genoa, Italy, 18–21 May 2015; pp. 1–6.
8. Pizzuti, L.; dos Santos Guimarães, C.; Iocca, E.G.; de Carvalho, P.H.S.; Martins, C.A. Continuous analysis of the acoustic marine noise: A graphic language approach. *Ocean Eng.* **2012**, *49*, 56–65. [CrossRef]
9. Hildebrand, J.A. Anthropogenic and natural sources of ambient noise in the ocean. *Mar. Ecol. Prog. Ser.* **2009**, *395*, 5–20. [CrossRef]
10. Van Walree, P. *Channel Sounding for Acoustic Communications: Techniques and Shallow-Water Examples*; Norwegian Defence Research Establishment (FFI), Technical Report FFI-Rapport; FFI: Kjeller, Norway, 2011; Volume 7.
11. Martin, R. Noise power spectral density estimation based on optimal smoothing and minimum statistics. *IEEE Trans. Speech Audio Process.* **2001**, *9*, 504–512. [CrossRef]

12. Cohen, I.; Berdugo, B. Noise estimation by minima controlled recursive averaging for robust speech enhancement. *IEEE Signal Process. Lett.* **2002**, *9*, 12–15. [CrossRef]
13. Cohen, I. Noise spectrum estimation in adverse environments: Improved minima controlled recursive averaging. *IEEE Trans. Speech Audio Process.* **2003**, *11*, 466–475. [CrossRef]
14. Hendriks, R.C.; Jensen, J.; Heusdens, R. Noise tracking using DFT domain subspace decompositions. *IEEE Trans. Audio Speech Lang. Process.* **2008**, *16*, 541–553. [CrossRef]
15. Taghia, J.; Taghia, J.; Mohammadiha, N.; Sang, J.; Bouse, V.; Martin, R. An evaluation of noise power spectral density estimation algorithms in adverse acoustic environments. In Proceedings of the IEEE International Conference on Acoustics, Speech and Signal Processing (ICASSP), Prague, Czech Republic, 22–27 May 2011; pp. 4640–4643.
16. Yu, G.; Mallat, S.; Bacry, E. Audio Denoising by Time-Frequency Block Thresholding. *IEEE Trans. Signal Process.* **2008**, *56*, 1830–1839. [CrossRef]
17. Courmontagne, P. The stochastic matched filter and its applications to detection and de-noising. In *Stochastic Control*; IntechOpen: London, UK, 2010.
18. Yu, D.; Deng, L.; Droppo, J.; Wu, J.; Gong, Y.; Acero, A. Robust speech recognition using a cepstral minimum-mean-square-error-motivated noise suppressor. *IEEE Trans. Audio Speech Lang. Process.* **2008**, *16*, 1061–1070.
19. Hu, Y.; Loizou, P.C. Speech enhancement based on wavelet thresholding the multitaper spectrum. *IEEE Trans. Speech Audio Process.* **2004**, *12*, 59–67. [CrossRef]
20. Moreaud, U.; Courmontagne, P.; Chaillan, F.; Mesquida, J.R. Performance assessment of noise reduction methods applied to underwater acoustic signals. In Proceedings of the OCEANS 2016 MTS/IEEE, Monterey, CA, USA, 19–23 September 2016; pp. 1–15.
21. Stein, M.C. Estimation of the Mean of a Multivariate Normal Distribution. *Ann. Stat.* **1981**, *9*, 1135–1151. [CrossRef]
22. Cai, T.T.; Zhou, H.H. A data-driven block thresholding approach to wavelet estimation. *Ann. Stat.* **2009**, *37*, 569–595. [CrossRef]
23. Li, D.Q.; Hallander, J.; Johansson, T. Predicting underwater radiated noise of a full scale ship with model testing and numerical methods. *Ocean Eng.* **2018**, *161*, 121–135. [CrossRef]
24. Coates, R.F. *Underwater Acoustic Systems*; Macmillan International Higher Education: London, UK, 1990.
25. Asolkar, P.; Das, A.; Gajre, S.; Joshi, Y. Comprehensive correlation of ocean ambient noise with sea surface parameters. *Ocean Eng.* **2017**, *138*, 170–178. [CrossRef]
26. Li, J.; White, P.R.; Bull, J.M.; Leighton, T.G. A noise impact assessment model for passive acoustic measurements of seabed gas fluxes. *Ocean Eng.* **2019**, *183*, 294–304. [CrossRef]
27. Porter, M.B. *The Bellhop Manual and User'S Guide: Preliminary Draft*; Technical Report; Heat, Light, and Sound Research, Inc.: La Jolla, CA, USA, 2011.

Journal of
Marine Science and Engineering

Article

An Improved Underwater Electric Field-Based Target Localization Combining Subspace Scanning Algorithm And Meta-EP PSO Algorithm

Wenjing Shang [1], Wei Xue [1], Yingsong Li [1,2], Xiangshang Wu [3] and Yidong Xu [1,*]

1 College of Information and Communication Engineering, Harbin Engineering University, Harbin 150001, China; shangwenjing@hrbeu.edu.cn (W.S.); xuewei@hrbeu.edu.cn (W.X.); liyingsong@hrbeu.edu.cn (Y.L.)
2 National Space Science Center, Chinese Academy of Sciences, Beijing 100190, China
3 Shanghai Mechanical & Electrical Engineering Research Institute, Shanghai 201109, China; wuxiangshang@foxmail.com
* Correspondence: xuyidong@hrbeu.edu.cn; Tel.: +86-132-6350-7375

Received: 13 February 2020; Accepted: 23 March 2020; Published: 26 March 2020

Abstract: In this paper, we propose an improved three-dimensional underwater electric field-based target localization method. This method combines the subspace scanning algorithm and the meta evolutionary programming (meta-EP) particle swarm optimization (PSO) algorithm. The subspace scanning algorithm is applied as the evaluation function of the electric field-based underwater target locating problem. The meta-EP PSO method is used to select M elite particles by the q-tournament selection method, which could effectively reduce the computational complexity of the three-dimensional underwater target localization. Moreover, the proposed meta-EP PSO optimization algorithm can avoid subspace scanning trapping into local minima. We also analyze the positioning performance of the uniform circular and cross-shaped electrodes arrays by using the subspace scanning algorithm combined with meta–EP PSO. According to the simulation, the calculation amount of the proposed algorithm is greatly reduced. Moreover, the positioning accuracy is effectively improved without changing the positioning accuracy and search speed.

Keywords: underwater localization; electric field; subspace scanning; meta-EP PSO

---

1. Introduction

Underwater target detection and estimation has a wide range of applications in marine salvage, marine exploration research, inspection of underwater facilities, underwater navigation and localization, and construction of an underwater environment [1–4]. However, due to the complexity of underwater environment, underwater target detection and estimation is still a challenging subject in theory and engineering practice [5–7]. In recent years, various underwater locating methods have been developed, including acoustic-, light-, and map-based locating methods [8–10]. At present, acoustic and optical imaging techniques are most commonly used in underwater target locating [11,12]. Acoustic signals have the advantage of less attenuation and longer underwater propagation distance than other methods. The underwater target positioning technology based on acoustic waves has provided a relatively complete theoretical system and has achieved considerable development [13,14]. However, the positioning performance of the acoustic method degrades due to specific factors, such as multipath effect, sonar scan angle, background noise, geomorphic structure complexity, and Doppler effect [15,16]. As the wavelength of the light is very short, the underwater positioning technology based on optical imaging has very high accuracy. Moreover, the situation is further complicated in shallow scenarios with rocks and sandbanks [17]. On the other hand, underwater imaging based on optical imaging cannot work in turbid water or environments with no light [18]. On the contrary, the underwater target locating methods based on electromagnetic fields can avoid these drawbacks [19]. Besides, the electromagnetic field-based localization methods do not suffer from the Doppler effect due to velocity higher than that of the sound waves, and they do not require transparent water [20]. Therefore, the localization methods based on the electromagnetic field have received great attention. Generally, the electromagnetic noise is extremely low and stable, especially in deep ocean environments because of the high conductivity of seawater [21]. The electromagnetic wave-based locating methods and the low-frequency electro-locating methods are two primary types of underwater locating methods based on the electromagnetic field. In [20,22], the locating methods based on the power path loss model of an electromagnetic wave propagating through seawater were proposed. Because of a small skin depth of a high-frequency signal in seawater, the power of the radio-frequency signal decreases dramatically, which makes it unsuitable for wide-range locating. The locating methods based on the quasistatic electric field have been widely studied [1,7,23,24], because they have lower path losses in seawater compared to the methods based on the high-frequency electromagnetic signals. The electric sense locating methods based on bionics show good performance in underwater avoidance, docking, and close-range object shape estimation in dark and turbid environments. However, electric sense active locating methods are not suitable for long-distance target locating because the electric field re-emitted by the target is usually much weaker than that of the source field. In Peng's work [25], the underwater target electric field locating method based on the coupling Cole–Cole model and finite element method is proposed. To locate the underwater target, one should move the electrode array and acquire the voltage in different point, limiting the application of the locating system. The Multiple Signal Classification (MUSIC) algorithm is a noniterative algorithm that can be used to create a space spectrum to locate an underwater electromagnetic source. In [26], a MUSIC-type algorithm was proposed for locating small inclusions buried in a half-space by measuring the scattering amplitude at a fixed frequency in a two-dimensional space. The locating method was based on the far-field theory. However, the far-field theory is not suitable for underwater target locating because high-frequency radiation waves cannot be transferred to a long distance. Therefore, in this paper, underwater target locating based on the quasi-static electric field for near-distance locating is introduced.

In this paper, we introduce the mixed polarization MUSIC algorithm for underwater localization. The mixed polarization MUSIC algorithm is different from the other MUSIC algorithms for radar, such as root-MUSIC and beamspace MUSIC: MP-MUSIC could deal with signal polarization, which is suitable for underwater electro-locating, allowing us to get the space position of a electric dipole without considering or solving the moment azimuth of the electric dipole, reducing the computation time [27,28]. The position of the target can be located via finding the minimum eigenvalue of the estimated gain matrix and the project matrix of the noise subspace by using the MP-MUSIC algorithm [29]. Searching for the solution to the proposed MUSIC algorithms denotes an optimization problem, so using a suitable optimization method can significantly reduce the calculation time. The evolutionary programming with a meta evolutionary programming (meta-EP) mutation algorithm and the particle swarm optimization algorithm are combined to develop a hybrid particle swarm algorithm for three-dimensional underwater target positioning. The simulations are conducted to validate the effectiveness of the proposed localization algorithm at different electrode configurations. The simulation results show that the proposed meta-EP particle swarm optimization (PSO) hybrid algorithm for searching an optimal solution to the localization algorithm has strong competitiveness in terms of accuracy and convergence speed.

2. Underwater Target Electro-Locating Method

2.1. Underwater Electric Field Forward Model

The schematic of the three-dimensional multi-electrode underwater electric field positioning is shown in Figure 1. In Figure 1, the electric dipole source is located at position $\mathbf{r}_p$, so the potential of the ith electrode can be calculated by

$$\varphi_{(i)} = k\frac{(\mathbf{r}_p - \mathbf{r}_i)^{\mathrm{T}}\mathbf{p}}{|\mathbf{r}_p - \mathbf{r}_i|^3} = \mathbf{g}\,(i)\,\mathbf{p}, \tag{1}$$

where $k = u_0/\,(4\pi)$ is a constant, $\mathbf{p}$ is the dipole moment, $\mathbf{r}_p$ is the location of electric dipole source, $\mathbf{r}_i$ is the location of the ith electrode, and $\mathbf{g}\,(i)$ is the gain vector of the ith receiving electric dipole located at position $\mathbf{r}_i$.

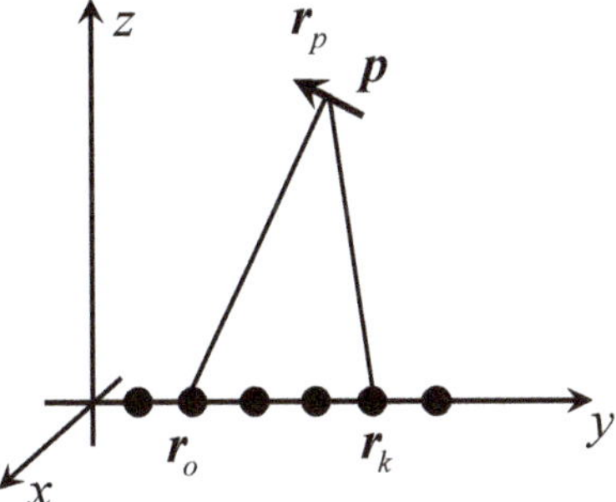

Figure 1. Schematic of the three-dimensional electric field-based multi-electrode underwater positioning.

The potential measured at different locations can be expressed as

$$\boldsymbol{\Psi} = \begin{bmatrix} \varphi_1 \\ \vdots \\ \varphi_m \end{bmatrix} = k\begin{bmatrix} \vdots \end{bmatrix}\mathbf{p} = k\begin{bmatrix} \mathbf{g}(1) \\ \vdots \\ \mathbf{g}(m) \end{bmatrix}\mathbf{p} = \mathbf{G}(\mathbf{r}_p)\mathbf{p}, \tag{2}$$

where $\mathbf{G}(\mathbf{r}_p)$ denotes the gain matrix. According to (2), the potential $\boldsymbol{\Psi}$ is linearly proportional to the dipole moment $\mathbf{p}$. Location parameter $\mathbf{r}_p$ in $\mathbf{G}(\mathbf{r}_p)$ is nonlinearly related to the potential $\boldsymbol{\Psi}$. Each column in $\mathbf{G}(\mathbf{r}_p)$ represents different dipole components of the same position. Therefore, for p-dipoles, according to the superposition theorem, the receiving potential can be expressed in the matrix form as

$$\boldsymbol{\Psi} = \begin{bmatrix} \mathbf{G}_1 & \cdots & \mathbf{G}_p \end{bmatrix}\begin{bmatrix} \mathbf{P}_1 \\ \vdots \\ \mathbf{P}_p \end{bmatrix}, \tag{3}$$

$$\mathbf{G}(\mathbf{r}) = \begin{bmatrix} \mathbf{G}_1 & \cdots & \mathbf{G}_p \end{bmatrix} = \begin{bmatrix} \mathbf{G}_1(\mathbf{r}_1) & \cdots & \mathbf{G}_p(\mathbf{r}_p) \end{bmatrix}, \tag{4}$$

$$\mathbf{T} = \begin{bmatrix} \mathbf{P}_1 \\ \vdots \\ \mathbf{P}_p \end{bmatrix}. \tag{5}$$

Equation (3) can be rewritten as $\boldsymbol{\Psi} = \mathbf{G}\,(\mathbf{r})\,\mathbf{T}$, where $\mathbf{G}_i(\mathbf{r}_i)$ denotes the gain matrix formulated by the ith dipole located at position $\mathbf{r}_i$, the receiving potential $\boldsymbol{\Psi}$ is a column vector with a size of $m \times 1$, $\mathbf{G}(\mathbf{r})$ is a matrix with a size of $m \times 3p$, and $\mathbf{T}$ is a column vector with a size of $3p \times 1$. Considering that the current intensity of electric dipole changes with time and that its position does not change, Equation (5) can be rewritten as

$$\mathbf{T} = \begin{bmatrix} \mathbf{M}_1 \begin{bmatrix} \mathbf{S}_1(1) & \cdots & \mathbf{S}_1(n) \end{bmatrix} \\ \vdots \\ \mathbf{M}_p \begin{bmatrix} \mathbf{S}_p(1) & \cdots & \mathbf{S}_p(n) \end{bmatrix} \end{bmatrix} = \begin{bmatrix} \mathbf{M}_1 & & 0 \\ & \ddots & \\ 0 & & \mathbf{M}_p \end{bmatrix} \begin{bmatrix} \mathbf{S}_1(1) & \cdots & \mathbf{S}_1(n) \\ \vdots & \ddots & \vdots \\ \mathbf{S}_p(1) & \cdots & \mathbf{S}_p(n) \end{bmatrix}, \tag{6}$$

where $\mathbf{M}_i$ represents the unit dipole moment of the ith electric dipole and $\mathbf{S}_i(j)$ denotes the amplitude of the ith electric dipole at time j. Therefore, Equation (3) can be expressed as

$$\begin{bmatrix} \varphi(1, 1) & \cdots & \varphi(1, n) \\ \vdots & \ddots & \vdots \\ \varphi(m, 1) & \cdots & \varphi(m, n) \end{bmatrix} = \begin{bmatrix} \mathbf{G}_1 & \cdots & \mathbf{G}_p \end{bmatrix} \begin{bmatrix} \mathbf{M}_1 & & 0 \\ & \ddots & \\ 0 & & \mathbf{M}_p \end{bmatrix} \begin{bmatrix} \mathbf{S}_1(1) & \cdots & \mathbf{S}_1(n) \\ \vdots & \ddots & \vdots \\ \mathbf{S}_p(1) & \cdots & \mathbf{S}_p(n) \end{bmatrix}. \tag{7}$$

Equation (7) can also be abbreviated as

$$\mathbf{\Psi} = \mathbf{GMS} = (\mathbf{GM})\,\mathbf{S} = \mathbf{HS}, \tag{8}$$

where $\mathbf{G}$ consists of p electric dipoles with a unit dipole moment and m receiving electrodes array, which forms a $m \times 3p$ matrix. The $3p \times p$ diagonal matrix $\mathbf{M}$ consists of p unit dipoles' moments with constant pointing. The dipole moment intensity matrix $\mathbf{S}$ has a dimension of $p \times n$; $\mathbf{H} = \begin{bmatrix} \mathbf{H}_1 & \cdots & \mathbf{H}_p \end{bmatrix} = \mathbf{GM}$, each column of $\mathbf{H}$ contains all the information about an electric dipole.

The electric field positioning can be considered as solving the minimum problem defined by

$$\mathbf{J}_f(i) = \lambda_{\min}\{\mathbf{U}_{\mathbf{G}_i}^{\mathrm{T}} \mathbf{P}^{\perp} \mathbf{U}_{\mathbf{G}_i}\}, \tag{9}$$

where $\lambda_{\min}\{\cdot\}$ denotes the minimum solution to the expression given in the curly brackets. Therefore, no special solution is required to make the minimum, and only the minimum eigenvalue related to the dipole moment needs to be calculated. The subspace scanning algorithm searches for possible locations of targets in a three-dimensional space. Accordingly, by finding the global minimum eigenvalue by eigenvalue decomposition, the target positioning in a three-dimensional space can be achieved.

2.2. Improved Three-Dimensional Subspace Scanning and Positioning Algorithm

In the three-dimensional underwater electric field-based target locating, it is necessary to obtain the received voltage data matrix using the receiving electrode array. The acquired data is given by

$$\mathbf{\Psi} = \mathbf{HS} + \mathbf{N}, \tag{10}$$

In Equation (10), the additive noise matrix $\mathbf{N}$ is assumed to be zero mean with the covariance of $E\{\mathbf{NN}^{\mathrm{T}}\} = \sigma_N{}^2\mathbf{I}$, where $E\{\cdot\}$ denotes the expected value of the argument, $\mathbf{H}$ denotes the gain matrix with a size of $(m < r)$, and $\mathbf{S}$ denotes a matrix of a size $r \times n(r < n)$. The expected value of the matrix outer product $\mathbf{R}_{\mathbf{\Psi\Psi}} = E\left\{\mathbf{\Psi\Psi}^{\mathrm{T}}\right\}$ can be represented under the zero-mean white noise assumption as follows,

$$\mathbf{R}_{\mathbf{\Psi\Psi}} = E\left\{[\mathbf{HS} + \mathbf{N}]\,[\mathbf{HS} + \mathbf{N}]^{\mathrm{T}}\right\} = \mathbf{HR}_{\mathbf{S}}\mathbf{H}^{\mathrm{T}} + \sigma_N{}^2\mathbf{I}, \tag{11}$$

where $\mathbf{R}_{\mathbf{S}} = E\{\mathbf{SS}^{\mathrm{T}}\}$, and $\mathbf{R}_{\mathbf{\Psi\Psi}}$ can be decomposed as

$$\mathbf{R}_{\mathbf{\Psi\Psi}} = \mathbf{U\Sigma U}^{\mathrm{T}} = \begin{bmatrix} \mathbf{U}_{\mathbf{S}} & \mathbf{U}_{\mathbf{N}} \end{bmatrix} \begin{bmatrix} \mathbf{\Sigma}_{\mathbf{S}} & \\ & \mathbf{\Sigma}_{\mathbf{N}} \end{bmatrix} \begin{bmatrix} \mathbf{U}_{\mathbf{S}} & \mathbf{U}_{\mathbf{N}} \end{bmatrix}^{\mathrm{T}}. \tag{12}$$

In Equation (12), the signal subspace $\mathbf{U_S}$ represents the vector space spanned by r eigenvectors corresponding to maximum eigenvalues. The remainder of $n-r$ eigenvector composes the noise subspace $\mathbf{U_N}$. Thus, Equation (9) can be rewritten as $\mathbf{J}_f(i) = \lambda_{\min}\left\{\mathbf{U}_{\mathbf{G}_i}^{\mathrm{T}}\mathbf{U_N}\mathbf{U_N^T}\mathbf{U}_{\mathbf{G}_i}\right\}$. The steps of the underwater target localization based on the subspace scanning algorithm are as follows.

- **Step 1:** Obtain measured voltage data using the receiving electrode array $\mathbf{\Psi}$.
- **Step 2:** Use Equation (11) to construct the corresponding covariance matrix $\mathbf{R}_{\mathbf{\Psi\Psi}}$.
- **Step 3:** Perform the eigenvalue decomposition on $\mathbf{R}_{\mathbf{\Psi\Psi}}$, and calculate the orthogonal projection matrix of the signal subspace $\mathbf{P}^{\perp}=\mathbf{U_N}\mathbf{U_N^T}$.
- **Step 4:** Scan each possible point r_i in a three-dimensional positioning area, calculate its gain vector $\mathbf{G}_i$, perform the singular value decomposition (SVD) operation to obtain the corresponding value $\mathbf{U}_{\mathbf{G}_i}$, evaluate each eigenvalue $\lambda_{\min}(\mathbf{U}_{\mathbf{G}_i}\mathbf{P}^{\perp}\mathbf{U}_{\mathbf{G}_i}^{\mathrm{T}})$, search first for the global minimum eigenvalue, and then the estimated point corresponding to the eigenvalue. The target position is estimated by the subspace scanning algorithm.

The proposed algorithm performs the eigenvalue decomposition operation on a gain matrix $\mathbf{G}_i$ at each possible position in the space and evaluates the corresponding singular value during the target positioning process in a three-dimensional space. The positioning process is computationally expensive. Assume a three-dimensional space 1 m × 1 m × 1 m, where the positioning area is divided using a 1-cm grid. To complete the scanning and positioning processes, it is necessary to perform 1,000,000 SVD and eigenvalue decomposition operations and calculate the corresponding evaluation process which is meshgrid scanning method. In the case of the same hardware platform configuration, usually, a larger number of calculations means a longer calculation time, and the positioning speed is slower.

With the aim to reduce the number of calculations of the subspace scanning algorithm in the positioning process, an improved subspace scanning algorithm based on a multi-step search operation and a simplex algorithm is proposed which is multi-step scanning method. The steps of the proposed target location algorithm are as follows.

- **Step 1:** Obtain measured voltage data using the receiving electrode array $\mathbf{\Psi}$.
- **Step 2:** Use Equation (11) to construct the corresponding covariance matrix $\mathbf{R}_{\mathbf{\Psi\Psi}}$.
- **Step 3:** Perform the eigenvalue decomposition on $\mathbf{R}_{\mathbf{\Psi\Psi}}$, and calculate the orthogonal projection matrix of the signal subspace $\mathbf{P}^{\perp}=\mathbf{U_N}\mathbf{U_N^T}$.
- **Step 4:** Scan each possible point $\mathbf{r}_i$ in a three-dimensional positioning area, calculate its gain vector $\mathbf{G}_i$, and perform the SVD operation to obtain the corresponding value $\mathbf{U}_{\mathbf{G}_i}$, then evaluate each eigenvalue $\lambda_{\min}\left(\mathbf{U}_{\mathbf{G}_i}\mathbf{P}^{\perp}\mathbf{U}_{\mathbf{G}_i}^{\mathrm{T}}\right)$, and search first for the global minimum eigenvalue, and then the estimated point corresponding to the eigenvalue. The target position is estimated by the subspace scanning algorithm.
- **Step 5:** Perform fine mesh division in the area near location r_{est}, and repeat **Step 4** to update the estimated location $\mathbf{r}_{est}$.
- **Step 6:** Repeat **Step 5** until the predefined minimum grid size is reached, and output the corresponding result $\mathbf{r}_{est-fin}$.
- **Step 7:** Use the simplex method to search for the initial point $\mathbf{r}_{est-fin}$; the obtained position represents the final target position estimated by the improved algorithm.

The multi-step scanning method can effectively reduce the calculation burden and improve the positioning speed. Assume a three-dimensional space 1m × 1m × 1m again. Suppose a 5 cm low-resolution coarse grid global scan is adopted, the corresponding spatial points are used as a starting point to perform a local grid fine-grained search with a resolution of 2 cm, 1 cm, 0.5 cm, 0.2 cm, and 0.1 cm in turn. The simplex method is used to search the local area for the initial point to obtain the final target position. The total number of scans is 48,000 + N (simplex), where N (simplex) denotes the number of searches performed by the simplex method, and the average value of N (simplex) of 100 is obtained by 1000 tests. Therefore, the multi-step scanning method for three-dimensional target positioning, compared with the meshgrid scanning method, can effectively reduce the number of calculations and can achieve positioning resolution of less than 0.1 cm. The multi-step scanning method can effectively increase the convergence speed, and thus improve the positioning speed.

In order to illustrate the effectiveness of the proposed positioning algorithm, assume that an electric dipole exists at the position (0.555, 0.555, 0.555) m with dipole moment orientation (1, 0, 0) A.m. A 100 Hz differential sine wave signal is loaded across the electric dipole. The uniform linear array, uniform circular array, and their modification array are commonly used in various applications [30,31]. The 8-channel uniform circular receiving electrodes with the circular radius R of 0.1 m are used for signal reception. The position information of the receiving electrodes is provided in Table 1, where electrode 9 that is at the center of the circle is set as a reference electrode, and the voltage is obtained by measuring the potential difference between it and other electrodes. The schematic diagram of the receiving electrode configuration is displayed in Figure 2. In Figure 2, the red dots represent the positive ends of the receiving electrodes, and the central black dot denotes the reference electrode. The received signal of electrode channels under the no-noise condition is presented in Figure 3.

The spatial spectrum image $L = 1/\lambda$ is drawn in the plane (x, y, 0.555) m. According to the analysis of the proposed algorithm, the dipole localization problem can be transformed into the problem of finding the minimum generalized eigenvalue, which is equivalent to finding the maximum of L. The bright spot position in Figure 4 has the largest value, and the corresponding point coordinate set is (0.555, 0.555, 0.555) m that consists of the positions predicted by the proposed positioning algorithm. The simulation results show that the dipole position can be predicted better by the algorithm under the no-noise condition, and the simulation output is consistent with the actual position.

Table 1. Position of receiving electrodes for uniform circular electrode configuration (unit: m).

Electrode	1	2	3	4	5	6	7	8	9
x	0.1	0.0707	0	−0.070	−0.1	−0.070	0	0.070	0
y	0	−0.0707	−0.1	−0.070	0	0.070	0.1	0.070	0
z	0	0	0	0	0	0	0	0	0

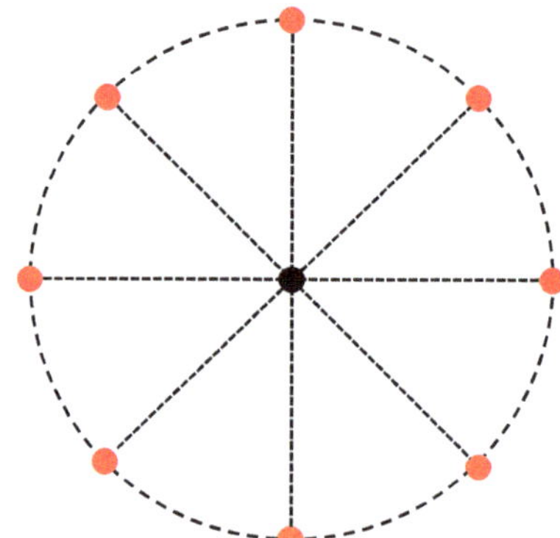

Figure 2. The uniform circular electrode configuration.

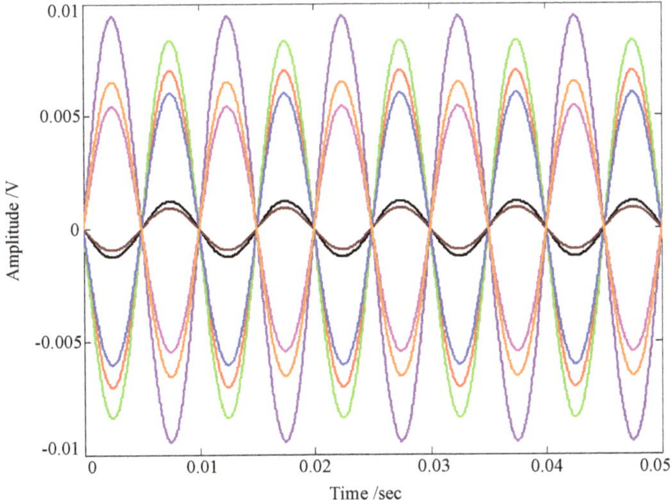

Figure 3. The received signal.

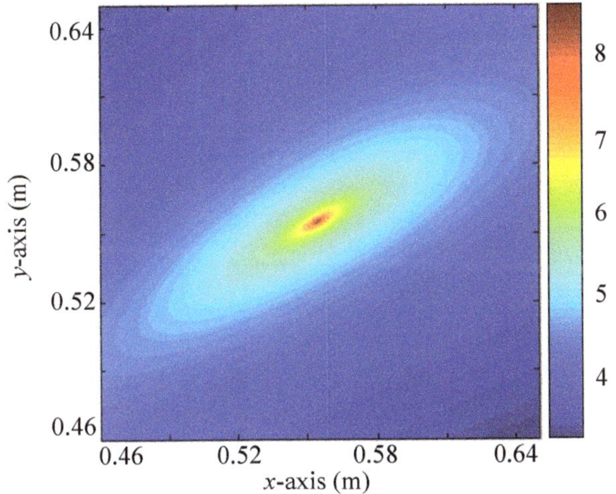

Figure 4. The spatial spectrum of in the plane (x, y, 0.555) m.

2.3. Electro-Location Based on PSO Algorithm

Although a multi-step search operation can effectively improve the search speed, it has been found that this method can fall into local extremes when performing the localization tests on some points. The PSO is a populated search method that employs a swarm of particles to probe the search space [32]. The PSO solves a problem by finding a population of candidate solutions, here the dubbed particles, and moving these particles around in the search-space following simple mathematical formulae over the particle's position and velocity; therefore, the PSO is relatively fast, simple, and can easily converge to the optimal solution. Therefore, the dipole localization has been determined by implementing the improved PSO procedures. A detailed description of the implemented optimization algorithm for solving the dipole localization problems is herein provided. To test the ability of meta-EP PSO for underwater dipole localization, we conducted the simulation experiments and compared the proposed algorithm with other versions of the PSO algorithm.

1) Original PSO algorithm

Each particle is treated as a point in a D-dimensional space. Suppose the ith particle is represented as $\mathbf{x}_i = (x_{i1}, x_{i2}, \cdots, x_{iD})$. The change rate of the velocity of the ith particle is represented as $\mathbf{v}_i = (v_{i1}, v_{i2}, \cdots, v_{iD})$. In the PSO algorithm, initially, a population of particles is randomly generated. The population update rules of the PSO algorithm at every iteration step are described as follows,

$$\begin{cases} \mathbf{v}_i^{t+1} = \mathbf{v}_i^t + c_1 \text{Rand}()(\mathbf{p}_i - \mathbf{x}_i^t) + c_2 \text{Rand}()(\mathbf{p}_g - \mathbf{x}_g^t) \\ \mathbf{x}_i^{t+1} = \mathbf{x}_i^t + \mathbf{v}_i^{t+1} \end{cases}, \tag{13}$$

where c_1 and c_2 denote the constants of canonical PSO; t represents the time step; Rand() stands for the random function in the range $[0, 1]$; and $\mathbf{p}_i$ and $\mathbf{p}_g$ denote the global best position and the personal best position of a particle, respectively.

2) Standard PSO (SPSO) algorithm

Shi and Eberhart [33] introduced an inertia weight w to improve the PSO accuracy by damping the velocities over time, allowing the swarm to converge with higher precision . By integrating w into the PSO algorithm, the velocity is updated by

$$\begin{cases} \mathbf{v}_i^{t+1} = w\mathbf{v}_i^t + c_1 \text{Rand}()(\mathbf{p}_i - \mathbf{x}_i^t) + c_2 \text{Rand}()(\mathbf{p}_g - \mathbf{x}_g^t) \\ \mathbf{x}_i^{t+1} = \mathbf{x}_i^t + \mathbf{v}_i^{t+1} \end{cases}, \tag{14}$$

A proper selection of the inertia weight ensures balance between the exploration and exploitation, where exploration represents the ability to test various regions in the problem space in order to achieve a good optimum, preferably the global one, and exploitation represents the ability to concentrate the search around a promising candidate solution in order to locate the optimum precisely. The choice of w defines how much the particle's current speed inherits. The more the particle inherits the current speed, the greater the global optimization ability, and the smaller the local search ability will be. Generally, fixed weight configuration and dynamic weight configuration are the two most common choices. According to the work in [32], the acceleration constants c_1 and c_2 can adjust and change the maximum step size of particles in time so that the particles can move in the direction of the best position of themselves. If the acceleration constants c_1 and c_2 are both equal to zero, the particles will move at the current speed until the boundary. In this case, the optimization process can be performed only in a limited range, which affects the algorithm performance. If the acceleration constant c_1 is set to be zero, it is a "social" model. The particles lack cognitive ability and rely only on the group experience. In this case, the algorithm converges quickly, but it can easily fall into a local optimum. On the other hand, when the acceleration constant c_2 is set to zero, it is a "cognitive" model. Particles cannot share socially, and rely only on their experience. In this case, it is difficult for the algorithm to find the global optimal value. Experiments have shown that there were no absolute optimal parameters, and it is necessary to determine appropriate parameters for each problem to obtain good convergence performance and robustness. Normally, the following values are used, $c_1 = c_2 = 2$[32].

3)Proposed meta-EP PSO algorithm

As the underwater target locating represents a nonconvex optimization problem, the parameter selection for a specific problem is not straightforward. As mentioned previously, the PSO algorithm has a risk of trapping into local minima and losing the exploration-exploitation ability. Thus, to overcome these shortcomings, an improved PSO algorithm that combines the movement update of the property of the canonical PSO algorithm with the meta-EP mutation characteristic is proposed.

In the proposed algorithm, M particles are selected among the swarm population by the q-tournament selection method [34]. Then, the selected elite particles are evolved using the meta-EP mutation and q-tournament selection of the EP [35]. The meta-EP mutation can be expressed as

$$\begin{cases} \mathbf{x}_i' = \mathbf{x}_i + \sqrt{\sigma_i} N_i(0,1) \\ \sigma_i' = \sigma_i + \sqrt{\alpha\sigma_i} N_i(0,1) \end{cases}, \tag{15}$$

where $\mathbf{x}_i$ denotes the position and σ_i denotes the standard deviation of Gaussian mutations. A single offspring $(\mathbf{x}_i', \sigma_i')$ is generated by parent particle $(\mathbf{x}_i, \sigma_i)$, where $N_i(0,1)$ indicates that the random number is generated for each iteration; α denotes an exogenous parameter ensuring that σ_i tends to remain positive.

By evaluating the fitness value of particles, the global best position is determined. According to the global best position, the nearest elite position, the personal best position, velocity, and position of a particle are updated in the next iteration using the following relations,

$$\begin{cases} \mathbf{v}_i^{t+1} = w\mathbf{v}_i^t + c_p Randp()(\mathbf{p}_i - \mathbf{x}_i^t) + c_g Randg()(\mathbf{p}_g - \mathbf{x}_g^t) + c_n Randn()(\mathbf{p}_e - \mathbf{x}_g^t) \\ \mathbf{x}_i^{t+1} = \mathbf{x}_i^t + \mathbf{v}_i^{t+1} \end{cases} \tag{16}$$

where c_g, c_n, and c_p denote the constant of the global best, the constant of the nearest elite, and the constant of the personal best, respectively; $Randp()$,$Randg()$,$Randn()$ represent random functions in the range [0, 1]. The proposed meta-EP algorithm for searching the position of a target includes the following steps.

Step 1: Initialize the positions of N particles, and evaluate the fitness values of all the particles.

Step 2: Select M elite particles by the q-tournament selection method.

Step 3: Evolve the elite particles by the EP and Equation (15).

Step 4: Evaluate the fitness values of the particles and determine the global best position.

Step 5: Determine the global best position, the nearest elite position, and the personal best position, and update the velocity and position of a particle according to (16).

Step 6: If the termination conditions are not satisfied, go to **Step 2**; otherwise, output the global best position.

2.4. Proposed Meta-EP PSO Algorithm for Underwater Dipole Localization

In this paper, the improved three-dimension subspace scanning and proposed meta-EP PSO algorithm is applied to underwater target localization. First, the forward model, electrode configuration, parameters, and the fitness function of the PSO are determined.

The flowchart of the proposed localization algorithm is presented in Figure 5. One of the key issues in the proposed algorithm is finding a suitable mapping between the localization problem solution and the PSO particle. The proposed PSO algorithm is applied to searching the solution to $\lambda_{\min}(\mathbf{U}_{\mathbf{G}_i}\mathbf{P}^{\perp}\mathbf{U}_{\mathbf{G}_i}^{\mathrm{T}})$. The dimension of the search space D, that is, the number of the elements of one particle, is equal to the number of position parameters of dipoles. For the source model with one dipole, D is equal to three, and a representation of dipole position is expressed as (x, y, z). The individuals in the swarm are initialized by setting their positions and velocities randomly in the searching space. Then, the velocity and position of particles are updated in each iteration. The optimization iteration is terminated when the pre-defined maximum iteration number is reached.

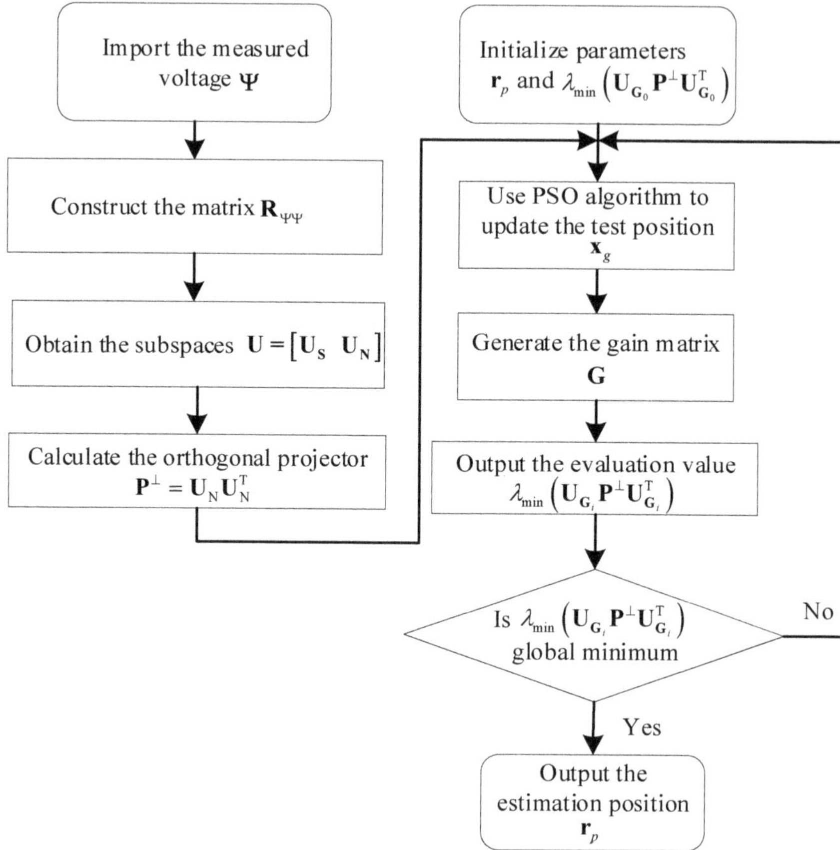

Figure 5. The flowchart of the proposed localization algorithm.

3. Numerical Simulations

In order to test the feasibility of the proposed PSO algorithm in the underwater target localizing, two simulation experiments were conducted to evaluate the positioning performance. The uniform circular electrode and cross electrode configurations are adopted as receiving electrode configurations. The schematic diagrams of the two common receiving electrode configurations are given in Figure 6a,b, where the red dots represent the positive receiving channels of the electrode channels, and the black dots represent the negative receiving channels of the electrode channels. In the two simulations, the radius R of the uniform circular receiving electrode was 0.1 m, and the electric dipole target was set in the plane xOy. The distance between the center of the receiving array and the target was r, and the electric dipole moment was $(1, 0, 0)$ m. At the signal-to-noise ratio of 40 dB, tests were performed 1000 times on each point, respectively, and the root mean square (RMS) error was calculated for each point at 0.5 m, 1 m, 2 m, 3 m, and 4 m, respectively.

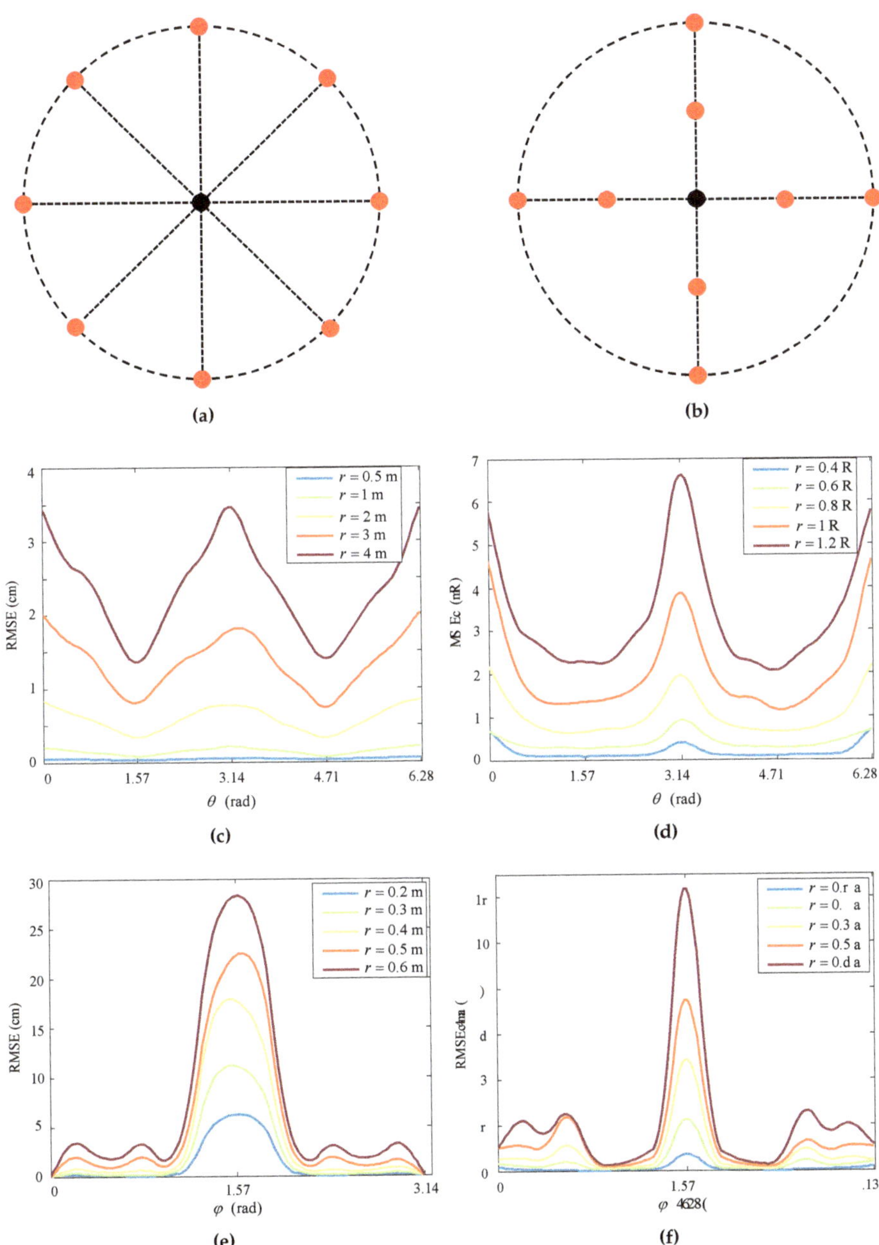

Figure 6. (**a**) The uniform circular electrode configuration; (**b**) The cross-shaped electrode configuration (**c**) Positioning error of uniform circular electrode configuration in the plane xOy; (**d**) Positioning error of the cross-shaped electrode configuration in the plane xOy; (**e**) Positioning error of uniform circular electrode configuration in the plane xOz; (**f**) Positioning error of the cross-shaped electrode configuration in the plane xOz.

3.1. Locating Performance of Uniform Circular Electrode Configuration

The electrode positions in the uniform circular receiving electrode configuration were the same as Table 1. The positioning performances at different positions in the planes xOy and plane xOz were studied, and the corresponding test results are shown in Figure 6c,e, respectively.

As presented in Figure 6c, (1) as the positioning distance increased in the plane xOy, the positioning error also increased; (2) at the same positioning distance r, the positioning error showed a certain regularity with the change in the deflection angle θ, namely, in the range of deflection angle from $(0, \pi/2)$ to $(\pi, 3\pi/2)$, the positioning error decreased with the deflection angle. On the other hand, in the range of the deflection angle from $(\pi/2, \pi)$ to $(3\pi/2, 2\pi)$, the positioning error increased with the deflection angle; (3) the locating system had blind points at the deflection angle of zero and π due to the symmetry of acquainted data—the received electrode voltage values of channels 2, 3, and 4 were, respectively, equal to that of channels 8, 7, and 6. For all the other points, the difference in the received voltage between the electrodes was small.

Similarly, in the case of a uniform circular electrode configuration, the positioning performance at different positions in the plane xOz was also studied, and the results are shown in Figure 6e. As shown in Figure 6e, the blind points occur at the norm direction of plane xOz because the signal intensity received by electrode channels 3 and 7 was equal to zero, whereas the elevation angle was $\pi/2$. By comparing the results presented in Figure 6c with those presented in Figure 6e, it can be found that the uniform circular electrode configuration provided better locating performance in the plane xOy.

3.2. Locating Performance of Cross-Shape Electrode Configuration

In the cross-shaped receiving electrode configuration, the electrode positions were as given in Table 2. The positioning performance at different positions in the planes xOy and xOz were studied, and the results are shown in Figure 6d,f, respectively.

Table 2. Position of receiving electrodes for cross-shaped electrode configuration (unit: m).

Electrode	1	2	3	4	5	6	7	8	9
x	−0.1	−0.05	0.05	0.1	0	0	0	0	0
y	0	0	0	0	−0.1	−0.05	0.05	0.1	0
z	0	0	0	0	0	0	0	0	0

Based on the results presented in Figure 6d, a similar conclusion with that of the uniform circular electrode configuration in the plane xOy can be obtained. However, the cross-shaped electrode configuration showed worse locating performance in the plane xOy compared with the uniform circular electrode configuration. For instance, the minimum locating error of the cross-shaped electrode configuration was larger than 1.5 cm, whereas the maximum locating error of the uniform circular electrode configuration was less than 0.3 cm.

Similarly, the positioning performance of the cross-shaped receiving electrode configuration at different positions in the plane xOz was also studied. The test results are shown in Figure 6f. As can be seen in Figure 6f, the blind point occurred in the direction normal to the xOz plane. However, compared with the uniform circular electrode configuration, the cross-shaped electrode configuration provided better locating performance. By comparing the positioning performances of the uniform circular receiving electrode configuration and the cross-shaped receiving electrode configuration, the following conclusions were drawn.

(1) For both the uniform circular receiving electrode configuration and the cross-shaped receiving electrode configuration, the positioning performance in the plane xOy was better than that in the plane xOz when the subspace scanning algorithm was used to locate underwater targets.
(2) The positioning performance of the uniform circular receiving electrode configuration was better than that of the cross-shaped receiving electrode configuration in the plane xOy. Moreover,

the positioning performance of the cross-shaped receiving electrode configuration was better than that of the uniform circular receiving electrode configuration in the plane xOz.

(3) Both configurations had certain positioning blind spots in the spatial three-dimensional positioning process.

4. Simulation and Analysis of the Proposed Algorithm

In this section, a simulation model and a detailed study of the proposed algorithm in underwater target locating are provided. The receiving array consisted of eight equidistant electrodes in the loop insulator framework. The positions of electrodes are shown in Table 1, and the radius of the loop insulator framework was 0.1 m. In the simulation, the dipole was placed at the position of $(0.555, 0.555, 0.555)$ m with the current moment of $(1, 0, 0)$ A·m.

The proposed meta-EP PSO algorithm was compared with the canonical PSO and SPSO algorithms. In order to ensure an objective comparison, the locating error was defined as

$$\mathrm{LE} = \sqrt{(x_{est} - x_o)^2 + (y_{est} - y_o)^2 + (z_{est} - z_o)^2}, \tag{17}$$

where $(x_{est}, y_{est}, z_{est})$ denoted the position estimated by an algorithm, and (x_o, y_o, z_o) denoted the actual dipole position.

The configuration parameters of the PSO, SPSO, and meta-EP PSO algorithms are given in Table 3. According to [32], in the fixed-weight configuration, the inertia weight w is commonly in the interval $[0.8, 1.2]$. Therefore, the dynamic weight configuration was used, where gradually decreased from 0.9 to 0.4. Accordingly, particles had different development and exploration capabilities at different stages of evolution. In the comparison, the population size was set to 30 and $c_1 = c_2 = 2$.

Table 3. The configuration parameters of the PSO, SPSO, and meta-EP PSO algorithms.

Algorithm	w	c_1	c_2	c_3	Size
PSO	-	2	2	-	30
SPSO	$0.9 \sim 0.4$	2	2	-	30
meta-EP PSO	$0.9 \sim 0.4$	0.8	0.4	0.8	30

Table 4 gives the average test results for 100 tests with 200 iterations each. The locating error of one of the tests is presented in Figure 7, and the position estimation in each iteration of the meta-EP PSO algorithm is presented in Figure 8.

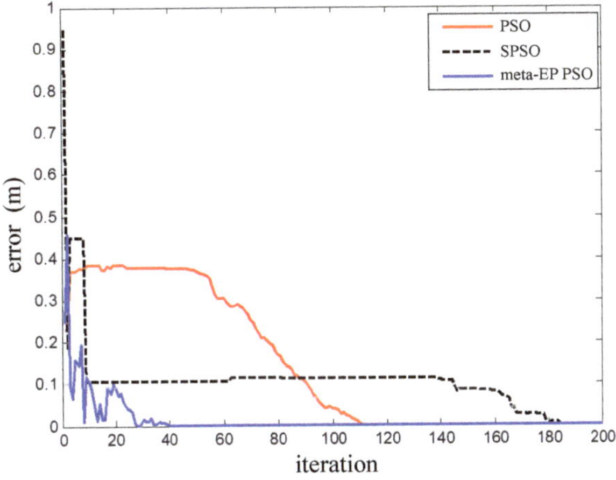

Figure 7. The locating error of one of the conducted tests.

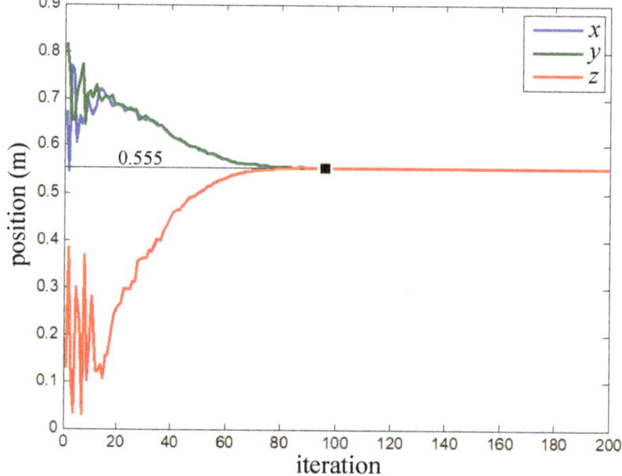

Figure 8. Position estimation in each iteration of the proposed meta-EP PSO algorithm.

Table 4. Test results of different algorithms.

Algorithm	PSO	SPSO	Meta-EP PSO
LE	0.545	0.356	0.006

As presented in Table 4, the PSO and SPSO algorithms converge to the local minimums during the positioning process, resulting in large positioning errors, which make them unsuitable for three-dimensional positioning scenarios. On the contrary, the proposed meta-EP PSO algorithm converged to the global minimum and provided the smallest positioning error among all the algorithms.

In order to study the computation of the meta-EP PSO further, we terminated the algorithm and recorded the number of iterations of the meta-EP PSO algorithm when the positioning error was less than 1 cm. The number of iterations of each test is shown in Figure 9.

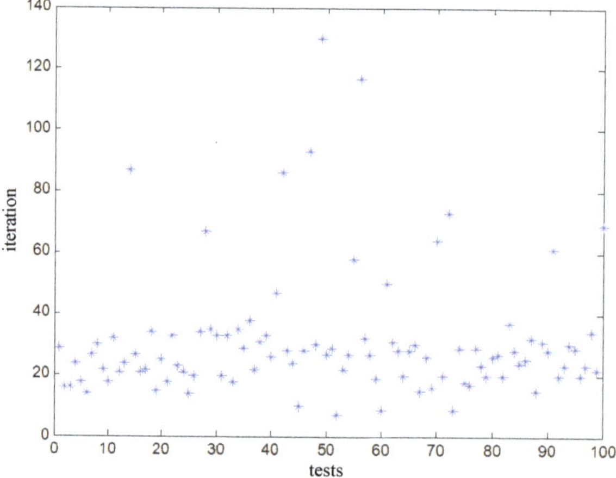

Figure 9. The number of iterations of each test.

As can be seen in Figure 9, most tests of the proposed algorithm terminated at up to 40 iterations. The average iteration number of the tests was 31.2, the maximum iteration number was 130, and the minimum iteration number was 7. The evolution of the algorithm in one iteration required the evaluation of all the particles in the population and the children generation. In this paper, the population size was set to 30, and the number of elite particles was set to 3. Thus, each iteration required at least 33 evaluations. Therefore, the maximum number of evaluation tests was 4290, and the average number of evaluation test was 1031.25. The average computation of the meta-EP PSO, meshgrid scanning method and the multi-step scanning are given in Table 5. Compared with the meshgrid scanning method of 1,000,000 times, the calculation amount of the proposed algorithm was greatly reduced to only 0.103% of the meshgrid scanning method. Similarly, the computation burden of meta-EP PSO is 2.14% of the multi-step method. The positioning accuracy was effectively improved without changing the positioning accuracy and search speed.

Table 5. Comparison of different methods.

Method	Evaluations
Meta-EP PSO	1031.25
Meshgrid scanning	1,000,000
Multi-step scanning	48,100

5. Conclusions

In this paper, we study the target locating in the underwater environment based on the electric field. The subspace scanning algorithm is applied as the evaluation function of the electric field-based underwater target locating problem. To find the global minimum of the evaluation function, the meta-EP PSO optimization algorithm is proposed. The meta-EP PSO method selects M elite particles by the q-tournament selection method, which could significantly speed up the convergence and avoid subspace scanning trapping into local minima. According to our simulations, the meta-EP PSO calculation burden is 0.10% of the meshgrid scanning method and 2.14% of the multi-step scanning method. The simulations show the meta-EP PSO provides more accurate locating performance, where the root mean square locating error is 0.006 m far smaller than the PSO and SPSO. Moreover, the meta-EP PSO shows fewer convergence steps compared with the PSO and SPSO. It takes the meta-EP PSO less than 40 generations to converge, whereas it takes totally 110 generations for PSO and 185 generations for SPSO. We also study the influence of the electrodes array on the locating performance. The uniform circular and the cross-shaped electrodes arrays are constructed. According to the simulations, we found the uniform circular electrodes array has better locating performance than that of the cross-shaped electrodes array in the plane xOy. However, the cross-shaped electrodes array shows better locating performance in the plane xOz. In our future work, we will optimize the electrode configurations to obtain a better locating performance.

Author Contributions: W.S. did the mathematical modeling and the simulations. W.S. also wrote the draft of the paper. Y.X. and X.W. contributed to the revisions and the discussion of the results. W.X. and Y.L. put forward to the idea and checked the simulation of this paper. All authors have read and agreed to the published version of the manuscript.

Funding: This work was supported in part by the Fundamental Research Funds for the Central Universities under Grant 3072019CFJ0802, in part by the Free Inquiry Projects of Fundamental Research Funds for the Central Universities, in part by the Fundamental Research Funds for the Central Universities under Grant HEUCFG201829, in part by the China Postdoctoral Science Foundation under Grant 2018M631911, and in part by the Heilongjiang Postdoctoral Fund under Grant LBH-Z18055.

Conflicts of Interest: The authors declare no conflicts of interest.

References

1. Lebastard, V.; Boyer, F.; Lanneau, S. Reactive underwater object inspection based on artificial electric sense. *Bioinspiration Biomim.* **2016**, *11*, 045003. [CrossRef] [PubMed]
2. Burguera, A. A novel approach to register sonar data for underwater robot localization. In Proceedings of the 2017 Intelligent Systems Conference (IntelliSys), London, UK, 7–8 September 2017; pp. 1034–1043.
3. Ol'shanskii, V.; Pavlov, D.; Volkov, S.; El'yashev, D. Electric fishes as a biological prototype of new technology. *Her. Russ. Acad. Sci.* **2009**, *79*, 64–77. [CrossRef]
4. Peralta, G.; Bonin-Font, F.; Caiti, A. Real-time Hash-based Loop Closure Detection in Underwater Multi-Session Visual SLAM. In Proceedings of the OCEANS 2019-Marseille, Marseille, France, 17–20 June 2019; pp. 1–7.
5. Bazeille, S.; Lebastard, V.; Lanneau, S.; Boyer, F. Model based object localization and shape estimation using electric sense on underwater robots. *IFAC-PapersOnLine* **2017**, *50*, 5047–5054. [CrossRef]
6. Boyer, F.; Lebastard, V.; Chevallereau, C.; Servagent, N. Underwater reflex navigation in confined environment based on electric sense. *IEEE Trans. Robot.* **2013**, *29*, 945–956. [CrossRef]
7. Bai, Y.; Snyder, J.B.; Peshkin, M.; MacIver, M.A. Finding and identifying simple objects underwater with active electrosense. *Int. J. Robot. Res.* **2015**, *34*, 1255–1277. [CrossRef]
8. Carroll, P.; Zhou, S.; Zhou, H.; Xu, X.; Cui, J.H.; Willett, P. Underwater localization and tracking of physical systems. *J. Electr. Comput. Eng.* **2012**. [CrossRef]
9. Lefort, R.; Real, G.; Drémeau, A. Direct regressions for underwater acoustic source localization in fluctuating oceans. *Appl. Acoust.* **2017**, *116*, 303–310. [CrossRef]
10. Burguera, A. Cluster-based Scan Matching for Robust Motion Estimation and Loop Closing. In Proceedings of the 2019 IEEE International Conference on Systems, Man and Cybernetics (SMC), Bari, Italy, 6–9 October 2019; pp. 2512–2517.
11. Al_Aboosi, Y.; Sha'ameri, A.Z. Experimental Multipath Delay Profile of Underwater Acoustic Communication Channel in Shallow Water. *Indones. J. Electr. Eng. Comput. Sci.* **2016**, *2*, 351–358. [CrossRef]
12. Marani, G.; Choi, S.K. Underwater target localization. *IEEE Robot. Autom. Mag.* **2010**, *17*, 64–70. [CrossRef]
13. Huang, Z.; Xu, J.; Gong, Z.; Wang, H.; Yan, Y. Source localization using deep neural networks in a shallow water environment. *J. Acoust. Soc. Am.* **2018**, *143*, 2922–2932. [CrossRef]
14. Xu, Y.; Guo, L.; Shang, W.; Li, Y. Underwater electro-location method based on improved matrix adaptation evolution strategy. *IEEE Access* **2018**, *6*, 39220–39232. [CrossRef]
15. Esmaiel, H.; Jiang, D. Multicarrier communication for underwater acoustic channel. *Int. J. Commun. Netw. Syst. Sci.* **2013**, *6*, 361.
16. Ebihara, T.; Leus, G. Doppler-resilient orthogonal signal-division multiplexing for underwater acoustic communication. *IEEE J. Ocean. Eng.* **2015**, *41*, 408–427.
17. Negre, P.L.; Bonin-Font, F.; Oliver, G. Cluster-based loop closing detection for underwater slam in feature-poor regions. In Proceedings of the 2016 IEEE International Conference on Robotics and Automation (ICRA), Stockholm, Sweden, 16–21 May 2016; pp. 2589–2595.
18. White, E.M.; Partridge, J.C.; Church, S.C. Ultraviolet dermal reflexion and mate choice in the guppy, Poecilia reticulata. *Anim. Behav.* **2003**, *65*, 693–700. [CrossRef]
19. Zazo, J.; Macua, S.V.; Zazo, S.; Pérez, M.; Pérez-Álvarez, I.; Jiménez, E.; Cardona, L.; Brito, J.H.; Quevedo, E. Underwater electromagnetic sensor networks, part II: Localization and network simulations. *Sensors* **2016**, *16*, 2176. [CrossRef] [PubMed]
20. Park, D.; Kwak, K.; Kim, J.; Chung, W.K. Underwater sensor network using received signal strength of electromagnetic waves. In Proceedings of the 2015 IEEE/RSJ International Conference on Intelligent Robots and Systems (IROS), Hamburg, Germany, 28 September–2 October 2015; pp. 1052–1057.
21. Shang, W.; Xue, W.; Xu, Y.; Geng, W. Undersea Target Reconstruction Based on Coupled Laplacian-of-Gaussian and Minimum Gradient Support Regularizations. *IEEE Access* **2019**, *7*, 171633–171647. [CrossRef]
22. Duecker, D.A.; Geist, A.R.; Hengeler, M.; Kreuzer, E.; Pick, M.A.; Rausch, V.; Solowjow, E. Embedded spherical localization for micro underwater vehicles based on attenuation of electro-magnetic carrier signals. *Sensors* **2017**, *17*, 959. [CrossRef]

23. Wang, K.; Do, K.D.; Cui, L. An underwater electrosensor for identifying objects of similar volume and aspect ratio using convolutional neural network. In Proceedings of the 2017 IEEE/RSJ International Conference on Intelligent Robots and Systems (IROS), Vancouver, BC, Canada, 24–28 September 2017; pp. 4963–4968.
24. Wang, K.; Do, K.D.; Cui, L. Underwater active electrosense: A scattering formulation and its application. *IEEE Trans. Robot.* **2017**, *33*, 1233–1241. [CrossRef]
25. Peng, J.; Wu, J. A numerical simulation model of the induce polarization: Ideal electric field coupling system for underwater active electrolocation method. *IEEE Trans. Appl. Supercond.* **2016**, *26*, 1–5. [CrossRef]
26. Ammari, H.; Iakovleva, E.; Lesselier, D. A MUSIC algorithm for locating small inclusions buried in a half-space from the scattering amplitude at a fixed frequency. *Multiscale Model. Simul.* **2005**, *3*, 597–628. [CrossRef]
27. Shirmehenji, F.; Nezhad, A.Z.; Firouzeh, Z.H. Object locating in anisotropic dielectric background using MUSIC algorithm. In Proceedings of the 2016 8th International Symposium on Telecommunications (IST), Tehran, Iran, 27–28 September 2016; pp. 396–400.
28. Xu, Y.; Shang, W.; Guo, L.; Qi, J.; Li, Y.; Xue, W. Active electro-location of objects in the underwater environment based on the mixed polarization multiple signal classification algorithm. *Sensors* **2018**, *18*, 554. [CrossRef] [PubMed]
29. Shahbazi, F.; Ziehe, A.; Nolte, G. Self-Consistent MUSIC algorithm to localize multiple sources in acoustic imaging. In Proceedings of the 4th Berlin Beamforming Conference, Berlin, Germany, 22–23 February 2012; pp. 22–23.
30. Shi, W.; Li, Y.; Zhao, L.; Liu, X. Controllable sparse antenna array for adaptive beamforming. *IEEE Access* **2019**, *7*, 6412–6423. [CrossRef]
31. Zhang, X.; Jiang, T.; Li, Y.; Zakharov, Y. A novel block sparse reconstruction method for DOA estimation with unknown mutual coupling. *IEEE Commun. Lett.* **2019**, *23*, 1845–1848. [CrossRef]
32. Marini, F.; Walczak, B. Particle swarm optimization (PSO). A tutorial. *Chemom. Intell. Lab. Syst.* **2015**, *149*, 153–165. [CrossRef]
33. Shi, Y.; Eberhart, R.C. Parameter selection in particle swarm optimization. In Proceedings of the International Conference on Evolutionary Programming, San Diego, CA, USA, 25–27 March 1998; pp. 591–600.
34. Lee, K.B.; Kim, J.H. Particle swarm optimization driven by evolving elite group. In Proceedings of the 2009 IEEE Congress on Evolutionary Computation, Trondheim, Norway, 18–21 May 2009; pp. 2114–2119.
35. Fogel, D.B. An introduction to simulated evolutionary optimization. *IEEE Trans. Neural Netw.* **1994**, *5*, 3–14. [CrossRef]

Letter

An Improved Sub-Array Adaptive Beamforming Technique Based on Multiple Sources of Errors

Zhuang Xie [1], Jiahua Zhu [2,*], Chongyi Fan [1] and Xiaotao Huang [1]

1 College of Electronic Science and Technology, National University of Defense Technology, Changsha 410073, China; xiezhuang18@nudt.edu.cn (Z.X.); chongyifan@nudt.edu.cn (C.F.); xthuang@nudt.edu.cn (X.H.)
2 College of Meteorology and Oceanography, National University of Defense Technology, Changsha 410073, China
* Correspondence: zhujiahua1019@nudt.edu.cn

Received: 27 August 2020; Accepted: 23 September 2020; Published: 28 September 2020

Abstract: In this paper, a new robust adaptive beamforming method is proposed in order to improve the robustness against steering vector (SV) mismatches that arise from multiple types of array errors. First, the sub-array technique is applied in order to obtain the decoupled sample covariance matrix (DSCM), in which the auxiliary sensors are selected to decouple the array. The decoupled interference-plus-noise covariance matrix (DINCM) is reconstructed with the estimated interference SV and maximum eigenvalue of the DSCM. Furthermore, the desired signal SV is estimated as the corresponding eigenvector determined by the correlation coefficients of the assumed SV and eigenvectors. Finally, the optimal weighting vector is obtained by combining the reconstructed DINCM and the estimated desired signal SV. Our simulation results show significant signal-to-interference-plus-noise ratio (SINR) enhancement of the proposed method over existing methods under multiple types of array errors.

Keywords: robust adaptive beamforming; steering vector mismatch; interference-plus-noise covariance matrix; array errors

---

1. Introduction

Adaptive beamforming has gained attention as an effective technique in array signal processing, due to its good target detection performance [1,2]. A Capon beamformer ensures the minimum output power under the premise of distortion-free reception from the desired signal direction, which is essentially equivalent to a minimum variance distortionless response (MVDR) beamformer [3,4], which is an optimal spatial filter, since it maximizes the output signal-to-interference-plus-noise ratio (SINR).

Although standard Capon beamformer is the theoretical optimal beamformer and has been widely applied for its good interference suppression ability, its performance drops sharply when there are mismatches between the assumed and real array model [5] due to various practical factors, such as inaccurate sensor positions [1], inconsistency of channels [6,7], and mutual coupling of antennas [8,9]. The above problems that are faced by the Capon beamformer are mainly divided into two categories: the mismatch of the desired steering vector (SV) and the involvment of the desired signal in the sample covariance matrix (SCM). The existence of the desired signal in the received snapshots significantly degraded the performance of the Capon beamformer, since the desired signal may be regarded as a interfernece and gets self-nulled [2]. The mismatch of the desired SV fails to steering the mainlobe towards the desired signal and, therefore, distorts the desired signal.

Numerous methods have been proposed to improve the robustness of Capon beamforming. Aside from the advantages of the robustness of the beamformer, the drawbacks of these algorithms are also obvious. The diagonal loading algorithm increases the robustness of the beamformer by adding a diagonal matrix on the sample covariance matrix in order to increase the noise power [10,11]. However, it is difficult to choose the appropriate diagonal loading factor. The eigenspace algorithm requires the specific number of interferences and it is able to provide satisfactory performance at some situations, but it is ineffective under low signal-to-noise ratio (SNR) conditions, since the desired signal subpace is swapped with the noise subpace [12–14]. The use of uncertainty set (US) algorithm is limited, as the size of the uncertain set is hard to determine and the desired signal is still involved in the SCM [15–24]. Interference-plus-noise covariance matrix (INCM) reconstruction-based algorithms have been shown to obtain excellent beamforming performance when the array manifold is accurately known [25–27], but they are not suitable for situations where an array of manifold mismatches exist [28].

The mutual coupling effect destorys the array structure in the SV and, therefore, affects the traditional methods. Ye et al. proposed a method where the mutual coupling effect could be mitigated by selecting middle array elements [9], but the presence of desired signal degrades its performance at high SNRs. Recently, the researchers combined the middle subarray technique and covariance matrix reconstruction technique in order to obtain the interfernce-noise covariance matrix in [29]. However, it should be noted that the method is based on the accurately known array structure, which is to say, the method is ineffective in the presence of other kinds of array errors, like sensor position errors and the gain-phase errors, since the real array structure is unavailable. In this paper, we improve the previous method in order to overcome the performance degradation tht arises from multiple types of array errors. Specifically, in terms of modification, our contributions are as follows.

- The characteristics of three different array error types and their influence on the recieved data are analyzed, a generalized signal model under the three kinds of errors is given.
- The middle array interference-plus-noise covariance matrix (INCM) is accurately reconstructed with estimated interference SV and power, which not only handles the problem of multiple types of array errors, but also mitigates the effect of the desired signal in the sample snapshots. The interference SVs are correctly estimated using the robust Capon beamforming (RCB) principle, as the SV mismatches that are due to the sensor position and gain-phase errors are relatively small. Furthermore, the estimated interference SVs are combined with the maximum eigenvalue of the decoupled sample covariance matrix (DSCM).
- The desired signal SV is estimated as the corresponding eigenvector of DSCM through the correlated projection process. The correlation coefficient of the SV and eigenvectors reaches the maximum when the eigenvector matches the SV.

The weighting vector is finally derived when combining the reconstructed middle array INCM and estimated desired signal SV. The proposed method is able to deal with multiple types of array errors and obtain superior SINR improvement. Throughout this paper, the superscripts T and H represent transpose and conjugate transpose, respectively. The notation $\mathrm{E}[\cdot]$ denotes the expectation operator and $\boldsymbol{I}$ stands for the unit matrix. $\odot$ is the Hadamard product. $[\cdot]^{-1}$ represents the matrix inversion operator.

2. Problem Formulation

2.1. Array Signal Model

Consider that there are $M+1$ narrowband signals $\{s_m(k)\}_{m=0}^{M}$ that impinge on the uniform linear array (ULA) of N array elements and they are uncorrelated with each other. That is to say,

$$\mathrm{E}\left[s_i s_j^H\right] = 0, \quad i \neq j \tag{1}$$

Assume that these signals arrive at the array with directions-of-arrivals (DOAs) $\{\theta_m\}_{m=0}^{M}$ and power $\{\sigma_m^2\}_{m=0}^{M}$. Let $s_0(k)$ represent the desired signal and the $\{s_m(k)\}_{m=1}^{M}$ are the interferences that are radiated by the farfield jammer devices. The received data in the k-th snapshot can be expressed as

$$\boldsymbol{x}(k) = \boldsymbol{A}\boldsymbol{S}(k) + \boldsymbol{n}(k) = s_0(k)\boldsymbol{a}(\theta_0) + \sum_{m=1}^{M} s_m(k)\boldsymbol{a}(\theta_m) + \boldsymbol{n}(k), \tag{2}$$

where $\boldsymbol{S}(k) = [s_0(k), s_1(k), \ldots, s_M(k)]^T$ denotes the echo signal vector and $\boldsymbol{n}(k)$ is an $N \times 1$ additive white Gaussian noise vector with power σ_n^2. The noise component is normal white Gaussian in the receiving channels, its model is assumed to be the same with traditional beamforming methods, since we mainly focus on the array errors in this paper. Further, $\boldsymbol{A} = [\boldsymbol{a}(\theta_0), \boldsymbol{a}(\theta_1), \ldots, \boldsymbol{a}(\theta_M)]$ stands for the steering matrix of the array, in which the m-th element is specifically given by $\boldsymbol{a}(\theta_m) = [1, b(\theta_m), \ldots, b(\theta_m)^{N-1}]^T$, where $b(\theta_m) = \exp(j2\pi d \sin\theta_m/\lambda)$, λ is the signal wavelength and d is the inter-element spacing. The $N \times 1$ weighting vector of the well-known Capon beamformer is given as:

$$\boldsymbol{w} = \frac{\boldsymbol{R}_{I+n}^{-1}\boldsymbol{a}(\theta_1)}{\boldsymbol{a}^H(\theta_1)\boldsymbol{R}_{I+n}^{-1}\boldsymbol{a}(\theta_1)}, \tag{3}$$

where $\boldsymbol{R}_{I+n} = \sum_{m=1}^{M} \sigma_m^2 \boldsymbol{a}(\theta_m)\boldsymbol{a}^H(\theta_m) + \sigma_n^2 \boldsymbol{I}_N$ is the INCM. In practice, the exact $\boldsymbol{R}_{I+n}$ is usually replaced by the sample covariance matrix (SCM), as $\hat{\boldsymbol{R}}_x \triangleq (1/K)\sum_{k=1}^{K} \boldsymbol{x}(k)\boldsymbol{x}^H(k)$, with K being the number of snapshots.

2.2. Array Error Model Analysis

In practice, array model errors essentially result in the mismatch of SV and they degrade the performance of traditional array signal processing algorithms, as shown in Figure 1. Under array errors, the ideal signal model Equation (2) is re-expressed as

$$\begin{aligned} \tilde{\boldsymbol{x}}(k) &= \tilde{\boldsymbol{A}}\boldsymbol{S}(k) + \boldsymbol{n}(k) \\ &= s_0(k)\tilde{\boldsymbol{a}}(\theta_0) + \sum_{m=1}^{M} s_m(k)\tilde{\boldsymbol{a}}(\theta_m) + \boldsymbol{n}(k) \end{aligned} \tag{4}$$

where $\tilde{\boldsymbol{A}} \triangleq f(\boldsymbol{A}, \Xi)$ is the actual steering matrix and Ξ is the matrix that identifies the array error. The new array structure is with the steering matrix $\tilde{\boldsymbol{A}}$ containing its array characteristics. In this section, the influences of three array error types are analyzed.

2.2.1. Mutual Coupling

Mutual coupling is an electromagnetic feature, where each sensor interacts with its neighbouring elements [8,9]. Let us define the mutual coupling length as P; that is, when considering the i-th element of the array, it couples with the $(i-P+1)$th, ..., $(i-1)$th, $(i+1)$th, ..., $(i+P-1)$th elements. The mutual coupling effect of the array can be expressed as a $M \times M$ symmetric Toeplitz matrix, as

$$\Xi_{MC} = \begin{bmatrix} 1 & c_1 & \cdots & c_{P-1} & \cdots & 0 \\ c_1 & 1 & c_1 & \cdots & \ddots & 0 \\ \vdots & c_1 & 1 & \ddots & \cdots & c_{P-1} \\ c_{P-1} & \cdots & \ddots & \ddots & c_1 & \vdots \\ 0 & \ddots & \cdots & c_1 & 1 & c_1 \\ 0 & \cdots & c_{P-1} & \cdots & c_1 & 1 \end{bmatrix}_{N \times N} \tag{5}$$

where c_p is the mutual coupling coefficient between the i-th and $(i \pm p)$th sensor. When the mutual coupling effect exists in the receiving array, Equation (4) is actually written as

$$\begin{aligned}\tilde{\boldsymbol{x}}(k) =&\tilde{\boldsymbol{A}}\boldsymbol{S}(k) + \boldsymbol{n}(k) \\ =&s_0(k)\tilde{\boldsymbol{a}}(\theta_0) + \sum_{m=1}^{M} s_m(k)\tilde{\boldsymbol{a}}(\theta_m) + \boldsymbol{n}(k) \\ =&s_0(k) \cdot (\Xi_{MC} \cdot \boldsymbol{a}(\theta_0)) + \sum_{m=0}^{M} s_m(k) \cdot (\Xi_{MC} \cdot \boldsymbol{a}(\theta_m)) + \boldsymbol{n}(k)\end{aligned} \tag{6}$$

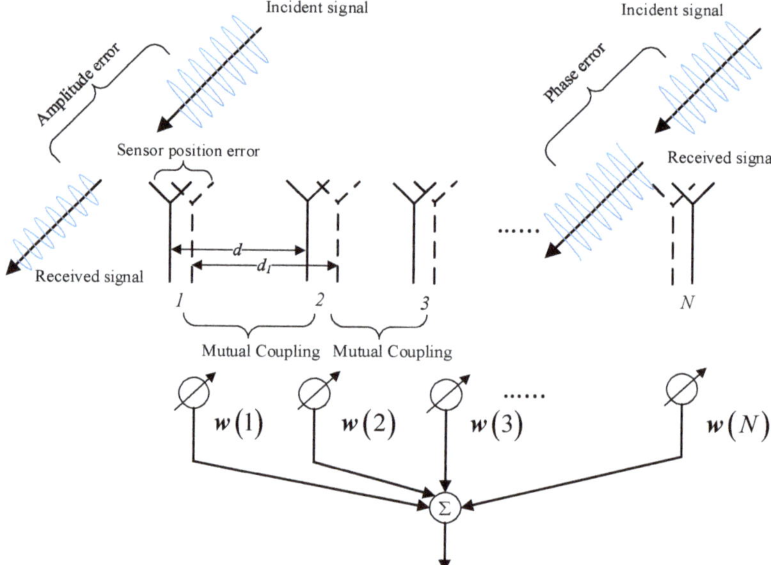

Figure 1. Receiving array in the presence of array errors.

2.2.2. Sensor Position Error

Realistic phenomena [1], such as sensor installation errors, measurement errors, and the instability of the antenna platform, inevitably induce sensor position errors. In general, the array element position error can be expressed, in matrix form, as

$$\Xi_{SP} = [\Delta\boldsymbol{a}^{\{1\}}, \Delta\boldsymbol{a}^{\{2\}}, \dots, \Delta\boldsymbol{a}^{\{M\}}] = \begin{bmatrix} \Delta a_1^{\{1\}} & \Delta a_2^{\{1\}} & \cdots & \Delta a_M^{\{1\}} \\ \Delta a_1^{\{2\}} & \Delta a_2^{\{2\}} & \cdots & \Delta a_M^{\{2\}} \\ \vdots & \vdots & \vdots & \vdots \\ \Delta a_1^{\{N\}} & \Delta a_2^{\{N\}} & \cdots & \Delta a_M^{\{N\}} \end{bmatrix} \tag{7}$$

where $\Delta\boldsymbol{a}^{\{m\}} = [\Delta a_m^{\{1\}}, \Delta a_m^{\{2\}}, \dots, \Delta a_m^{\{N\}}]^T$ stands for the array mismatch vector for the signal from direction θ_m. Specifically, its n-th element can be expressed as $\Delta a_m^{\{n\}} = \exp(j2\pi \sin\theta_m \Delta d_n/\lambda)$, where $\Delta d_n = \sum_{i=0}^{n-1} d_i - (n-1)d$, with d_0 set to 0 and where d_i represents the real spacing between the i-th sensor and the $(i+1)$th sensor. When the sensor position errors exist in the receiving array, Equation (4) is actually written as

$$\begin{aligned}\tilde{x}(k) =& \tilde{A}S(k) + n(k) \\ =& s_0(k)\tilde{a}(\theta_0) + \sum_{m=1}^{M} s_m(k)\tilde{a}(\theta_m) + n(k) \\ =& s_0(k) \cdot \left(\Delta a^{\{0\}} \odot a(\theta_0)\right) + \sum_{m=0}^{M} s_m(k) \cdot \left(\Delta a^{\{m\}} \odot a(\theta_m)\right) + n(k)\end{aligned} \quad (8)$$

2.2.3. Gain-Phase Error in Channel

Because of variations in time and temperature, the gain-phase characteristics of the receiving sensors change accordingly [6,7]. The gain-phase error can be characterized, by a diagonal matrix, as

$$\Xi_{GP} = \begin{bmatrix} \gamma_1 & 0 & \cdots & 0 \\ 0 & \gamma_2 & \cdots & 0 \\ 0 & 0 & \ddots & 0 \\ 0 & 0 & \cdots & \gamma_N \end{bmatrix} \quad (9)$$

where $\gamma_n = \alpha_n \exp(j\beta_n)$, and α_n and β_n are the additional gain-phase errors of the n-th channel. When the gain-phase error exists in the receiving array, Equation (4) is actually written as

$$\begin{aligned}\tilde{x}(k) =& \tilde{A}S(k) + n(k) \\ =& s_0(k)\tilde{a}(\theta_0) + \sum_{m=1}^{M} s_m(k)\tilde{a}(\theta_m) + n(k) \\ =& s_0(k) \cdot (\Xi_{GP} \cdot a(\theta_0)) + \sum_{m=0}^{M} s_m(k) \cdot (\Xi_{GP} \cdot a(\theta_m)) + n(k)\end{aligned} \quad (10)$$

When all three kinds of errors detailed above exist in the array, the actual steering matrix $\tilde{A}$ can be calculated, as

$$f\,(A, \Xi) = \Xi_{MC} \cdot \Xi_{GP} \cdot (\Xi_{SP} \odot A)\,. \quad (11)$$

From the above analysis, it can be seen that SV mismatches are due to the array errors, which severely degrade the performance of beamforming methods. Furthermore, the SV mismatches that arise from mutual coupling are far larger than those of sensor position and gain-phase errors.

3. Proposed Robust Adaptive Beamforming Method

In this section, we propose a new beamforming method to effectively suppress the interferences and noise in the presence of multiple types of array errors. The DINCM is accurately reconstructed based on the constructed DSCM, together with the estimated desired signal SV, in order to form the proposed beamformer. The detailed procedures are as follows:

3.1. DSCM Construction Based on Sub-Array

When the three types of array errors that are introduced above coexist in the array, then the actual steering matrix $\tilde{A}$ can be calculated as $f\,(A, \Xi) = \Xi_{MC} \cdot \Xi_{GP} \cdot (\Xi_{SP} \odot A)$. To begin with, the actual received data Equation (4) can be further modified as

$$\begin{aligned}\tilde{x}(k) =& \tilde{A}S(k) + n(k) \\ =& \Xi_{MC}\tilde{A}'S(k) + n(k)\end{aligned} \quad (12)$$

where $\tilde{A}' = \Xi_{GP} \cdot (\Xi_{SP} \odot A)$. The signal expression in Equation (12) can be viewed as an array with mutual coupling, with ideal steering matrix $\tilde{A}' = [\tilde{a}'(\theta_0), \tilde{a}'(\theta_1), \ldots, \tilde{a}'(\theta_M)]$. Its elements have the form $\tilde{a}'(\theta_m) = [\Delta b^{\{1\}}(\theta_m), \Delta b^{\{2\}}(\theta_m) \cdot b(\theta_m), \ldots, \Delta b^{\{N\}}(\theta_m) \cdot b(\theta_m)^{N-1}]^T$, where $\Delta b^{\{n\}}(\theta_m) = \alpha_n \cdot \Delta a_m^{\{n\}} \cdot \exp(j\beta_n)$. In order to mitigate the mutual coupling effect in the array $\tilde{Y}'$, the $N-2P+2$ sensors in the middle are chosen as the sub-array. In this sense, the $P-1$ sensors in the front and end are used as auxiliary sensors [9]. For convenience, we use N' to represent $N-2P+2$ in the rest of this paper. Therefore, the data of the sub-array is selected as

$$\bar{\boldsymbol{x}}(k) = \boldsymbol{\Gamma}\tilde{\boldsymbol{x}}(k) = \boldsymbol{\Gamma}\Xi_{MC}\tilde{\boldsymbol{A}}'\boldsymbol{S}(k) + \boldsymbol{\Gamma}\boldsymbol{n}(k), \tag{13}$$

where $\boldsymbol{\Gamma} = [\boldsymbol{O} \quad \boldsymbol{I}_{N'} \quad \boldsymbol{O}]$ is the selective matrix, $\boldsymbol{O}$ is an $N' \times (P-1)$ matrix with all elements being zero. If we use $\tilde{\boldsymbol{A}}''$ to denote $\boldsymbol{\Gamma}\Xi_{MC}\tilde{\boldsymbol{A}}'$, then $\tilde{\boldsymbol{A}}'' = [\tilde{\boldsymbol{a}}''(\theta_0), \tilde{\boldsymbol{a}}''(\theta_1), \ldots, \tilde{\boldsymbol{a}}''(\theta_M)]$, in which $\tilde{\boldsymbol{a}}''(\theta_m)$ is expressed as $\tilde{\boldsymbol{a}}''(\theta_m) = g(\theta_m) \cdot [\Delta b^{\{1\}}(\theta_m), \Delta b^{\{2\}}(\theta_m) \cdot b(\theta_m), \ldots, \Delta b^{\{N'\}}(\theta_m) \cdot b(\theta_m)^{N'-1}]^T$, where $g(\theta_m) = 1 + \sum_{i=1}^{P-1} c_{i+1}\left(b'(\theta_m)^{-i} + b'(\theta_m)^{i}\right)$ and $b'(\theta_m) = \Delta b^{\{2\}}(\theta_m) \cdot b(\theta_m)$. It can be shown that the mutual coupling effect in the data is eliminated by multiplying the original data with the selection matrix. Therefore, the DSCM is constructed, as $\hat{\boldsymbol{R}}_x \triangleq (1/K)\sum_{k=1}^{K} \bar{\boldsymbol{x}}(k)\bar{\boldsymbol{x}}^H(k)$.

3.2. Accurate DINCM Reconstruction

In [29], the researchers simply utilize the Capon spectrum to integrate in the interference region to reconstruct the INCM. However, this method is ineffective and it suffers severe performance degradation when multiple types of array errors exist, as shown in the simualtion part. To effectively form deep nulls in the interferences and noise, in this paper, we shall show how the DINCM is reconstructed in a improved way in order to achieve robustness to multiple type of errors. To begin with, the Capon spatial spectrum [30] is utilized to obtain an approximate estimate of the interference DOAs. The expression is given as

$$\hat{P}_{Capon}(\theta) = \frac{1}{\bar{\boldsymbol{a}}^H(\theta)\hat{\boldsymbol{R}}_x^{-1}\bar{\boldsymbol{a}}(\theta)}, \tag{14}$$

where $\bar{\boldsymbol{a}}(\theta_m) = [1, b(\theta_m), \ldots, b(\theta_m)^{N'-1}]^T$. By searching in the complement sector of the desired signal region, the DOAs of the searched peaks $\hat{\theta}_1, \hat{\theta}_2, \ldots, \hat{\theta}_M$ are utilized in order to obtain an approximate estimate of interference SV as $\bar{\boldsymbol{a}}(\hat{\theta}_1), \bar{\boldsymbol{a}}(\hat{\theta}_2), \ldots, \bar{\boldsymbol{a}}(\hat{\theta}_M)$. As the SV mismatches due to gain-phase and sensor position errors are relatively small, the accuracy can be enhanced by correcting the SVs with the RCB principle. The correction processing for the m-th SV can be performed by solving

$$\min_{\boldsymbol{a}_m^{(N')}} \boldsymbol{a}_m^{(N')H}\hat{\boldsymbol{R}}_x^{-1}\boldsymbol{a}_m^{(N')} \quad s.t. \|\boldsymbol{a}_m^{(N')} - \bar{\boldsymbol{a}}(\hat{\theta}_m)\|^2 \leq \epsilon, \tag{15}$$

where ϵ is the uncertainty level, which indicates the extent of SV mismatches. Therefore, after solving M problems, the corrected SVs $\left\{\boldsymbol{a}_m^{(N')}\right\}_{m=1}^{M}$ can be obtained. The solution of the m-th problem is given by $\boldsymbol{a}_m^{(N')} = \bar{\boldsymbol{a}}(\hat{\theta}_m) - \left(\boldsymbol{I}_{N'} + \delta\hat{\boldsymbol{R}}_x\right)^{-1}\bar{\boldsymbol{a}}(\hat{\theta}_m)$, where δ is the Lagrange multiplier and it can be calculated by solving $\|\left(\boldsymbol{I}_{N'} + \delta\hat{\boldsymbol{R}}_x\right)^{-1}\bar{\boldsymbol{a}}(\hat{\theta}_m)\|^2 = \epsilon$. On the other hand, the power of the interferences can be approximated by the corresponding eigenvalue divided by the array size [28]. If we denote the eigendecomposition of DSCM as $\hat{\boldsymbol{R}}_x = \sum_{n=1}^{N'} \hat{\lambda}_n\hat{\boldsymbol{u}}_n\hat{\boldsymbol{u}}_n^H$, where $\hat{\lambda}_1 \geq \hat{\lambda}_2 \geq \cdots \geq \hat{\lambda}_{N'}$ are eigenvalues that are arranged in a descending order and $\hat{\boldsymbol{u}}_n$ corresponds to $\hat{\lambda}_n$. Subsequently, in terms of the interference powers, they can be specifically estimated as

$$\sigma_m^2 \approx \frac{\hat{\lambda}_m}{N'}, \quad m = 1, \ldots, \tag{16}$$

In practice, in order to make sure that the interference powers are not underestimated, the maximum power is used for all of the interference powers. The following estimate is derived

$$\sigma_m^2 \approx \frac{\hat{\lambda}_1}{N'}, \quad m = 1, \ldots, \tag{17}$$

With the estimated power and SV of the interferences, the interference covariance matrix is reconstructed as

$$\hat{\hat{\boldsymbol{R}}}_{I+n} = \sum_{m=1}^{M} \frac{\hat{\lambda}_1}{N'} \boldsymbol{a}_m^{(N')} \boldsymbol{a}_m^{(N')H} \tag{18}$$

On the other hand, using the minimum eigenvalue $\hat{\lambda}_{N'}$ as the estimate of noise power [31], the noise covariance matrix is reconstructed as

$$\hat{\hat{\boldsymbol{R}}}_n = \hat{\lambda}_{N'} \boldsymbol{I}_{N'} \tag{19}$$

Combining the above processes, the following DINCM reconstruction expression can be derived

$$\hat{\hat{\boldsymbol{R}}}_{I+n} = \sum_{m=1}^{M} \frac{\hat{\lambda}_1}{N'} \boldsymbol{a}_m^{(N')} \boldsymbol{a}_m^{(N')H} + \hat{\lambda}_{N'} \boldsymbol{I}_{N'}. \tag{20}$$

3.3. Desired Signal SV Estimation

The SV of the desired signal can be replaced by the corresponding eigenvector, as the desired signal covariance matrix is rank one. The eigenvector that corresponds to the desired signal SV can be chosen by projecting the eigenvectors into the assumed SV (i.e., the correlation coefficient of the SV and the eigenvectors reaches the maximum when the eigenvector matches the SV) [32]. The correlation coefficient between the i-th eigenvector and assumed SV is defined as

$$cor(\hat{\boldsymbol{u}}_i, \bar{\boldsymbol{a}}(\theta_0)) = \frac{|\hat{\boldsymbol{u}}_i^H \bar{\boldsymbol{a}}(\theta_0)|}{\|\hat{\boldsymbol{u}}_i\| \|\bar{\boldsymbol{a}}(\theta_0)\|} \tag{21}$$

The correlation coefficient between $\hat{\boldsymbol{u}}_i$ and $\bar{\boldsymbol{a}}(\theta_0)$ reaches maximum when $\hat{\boldsymbol{u}}_i$ is the eigenvector that corresponds to the desired signal. Therefore, the desired signal SV is obtained as

$$\boldsymbol{a}_0^{(N')} = \sqrt{N'} \boldsymbol{u}_d, \tag{22}$$

where $\boldsymbol{u}_d$ is the solution to the problem

$$\max_{\hat{\boldsymbol{u}}_i} |\hat{\boldsymbol{u}}_i^H \bar{\boldsymbol{a}}(\theta_0)| \quad s.t. \quad 1 \leq i \leq M+1. \tag{23}$$

By replacing the theoretical DINCM and SV of desired signal with $\hat{\hat{\boldsymbol{R}}}_{I+n}$ and $\boldsymbol{a}_0^{(N')}$, the proposed beamformer is given as

$$\boldsymbol{w}_{PRAB} = \frac{\hat{\hat{\boldsymbol{R}}}_{I+n}^{-1} \boldsymbol{a}_0^{(N')}}{\boldsymbol{a}_0^{(N')H} \hat{\hat{\boldsymbol{R}}}_{I+n}^{-1} \boldsymbol{a}_0^{(N')}}. \tag{24}$$

By applying the weighting vector $\boldsymbol{w}_{PRAB}$ to the beamformer, the received data can be processed in order to effectively suppress the interference and noise. Specifically, the output of the beamformer at instant k is given as

$$y_{out}(k) = \boldsymbol{w}_{PRAB}^H \bar{\boldsymbol{x}}(k). \tag{25}$$

The main complexity of our proposed method lies in the interference SV estimation and DSCM eigendecomposition. Let us define J as the number of search points in the Capon spectrum, and then the computational complexity of the interference SV estimation and DSCM eigendecomposition are about

$\mathcal{O}\left(N'^3 + N'^2J + N'^2K\right)$ and $\mathcal{O}\left(N'^3\right)$, in terms of the number of flops, respectively. When considering the fact that $J > K > N'$, the above complexity actually becomes $\mathcal{O}\left(N'^2J\right)$. Therefore, the overall complexity of the proposed approach is about $\mathcal{O}\left(N'^2J\right)$. Algorithm 1 summarizes the proposed method. An additional flow chart figure of the proposed method is provided in Figure 2, where the application of sub-array technique is presented in a clearer way.

Algorithm 1 Steps of the proposed robust adaptive beamforming method

Part 1. DSCM construction based on sub-array

1. Decoupling the received data as $\bar{\boldsymbol{x}}(k) = \boldsymbol{\Gamma}\tilde{\boldsymbol{x}}(k) = \boldsymbol{\Gamma}\boldsymbol{\Xi}_{MC}\tilde{\boldsymbol{A}}'\boldsymbol{S}(k) + \boldsymbol{\Gamma}\boldsymbol{n}(k)$.

2. Constructing the DSCM as $\hat{\bar{\boldsymbol{R}}}_x \triangleq (1/K)\sum_{k=1}^{K}\bar{\boldsymbol{x}}(k)\bar{\boldsymbol{x}}^H(k)$.

Part 2. Accurate DINCM reconstruction

3. Obtaining approximate estimates of the interference SVs utilizing the Capon spectrum as $\bar{\boldsymbol{a}}(\hat{\theta}_1), \bar{\boldsymbol{a}}(\hat{\theta}_2), \ldots, \bar{\boldsymbol{a}}(\hat{\theta}_M)$.

4. Correcting the interference SVs with RCB principle and obtaining the corrected SVs $\left\{\boldsymbol{a}_m^{(N')}\right\}_{m=1}^{M}$.

5. Eigendecomposing the DSCM as $\hat{\bar{\boldsymbol{R}}}_x = \sum_{n=1}^{N'}\hat{\lambda}_n\hat{\boldsymbol{u}}_n\hat{\boldsymbol{u}}_n^H$ and reconstructing the DINCM as $\hat{\bar{\boldsymbol{R}}}_{I+n} = \sum_{m=1}^{M}\frac{\hat{\lambda}_1}{N'}\boldsymbol{a}_m^{(N')}\boldsymbol{a}_m^{(N')H} + \hat{\lambda}_{N'}\boldsymbol{I}_{N'}$.

Part 3. Desired signal SV estimation

6. Choosing out the eigenvector of $\hat{\bar{\boldsymbol{R}}}_x$ maximizes the projection into the assumed SV by solving $\max_{\hat{\boldsymbol{u}}_i}|\hat{\boldsymbol{u}}_i^H\bar{\boldsymbol{a}}(\theta_0)|\quad s.t.\quad 1 \le i \le M+1$.

7. Obtaining the desired signal SV as $\boldsymbol{a}_0^{(N')} = \sqrt{N'}\boldsymbol{u}_d$.

Final Calculating the weighting vector

$$w_{PRAB} = \frac{\hat{\bar{\boldsymbol{R}}}_{I+n}^{-1}\boldsymbol{a}_0^{(N')}}{\boldsymbol{a}_0^{(N')H}\hat{\bar{\boldsymbol{R}}}_{I+n}^{-1}\boldsymbol{a}_0^{(N')}}.$$

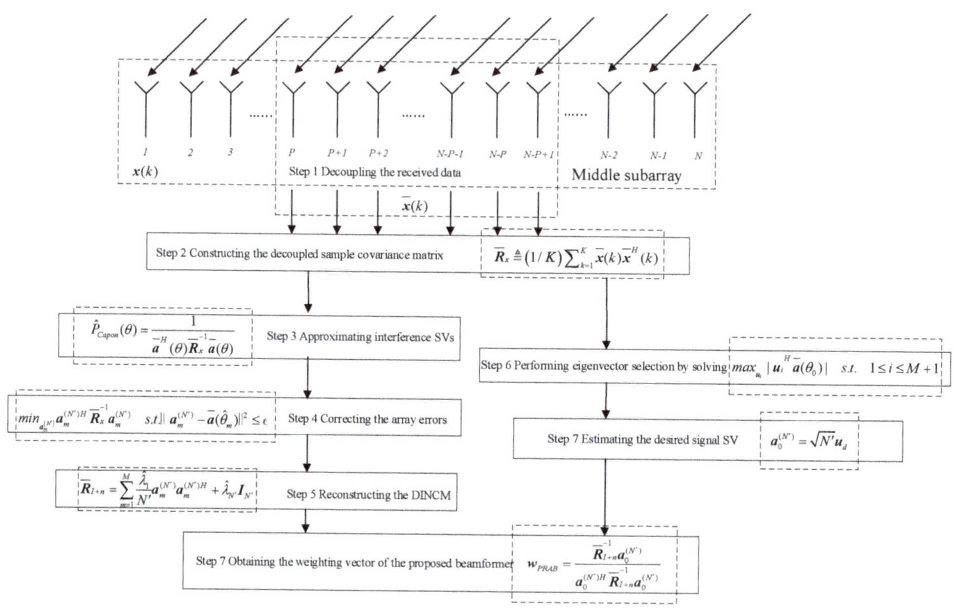

Figure 2. Flow chart of the proposed method.

4. Simulation Results

A ULA with 28 receiving sensors with half-wavelength inter-element spacing was deployed in the considered scenario. Two interferences were assumed from DOAs of -50° and 32° with INR 20 dB, while the desired signal was pointed with a DOA of 0°. The additive noise was set to be white Gaussian noise with unit variance.

The simulation considered all three types of array model errors. The sensor position error satisfied the normal distribution $N(d, (0.025)^2)$. The gain-phase error followed the normal distributions $N(1, 0.1^2)$ and $N(0, (5^\circ)^2)$, respectively. $P = 4$ mutual coupling was considered in the scene and $c_1 = 1.65e^{-j\pi/3}$, $c_2 = 2.35e^{j\pi/2}$, and $c_3 = 0.25e^{-j2\pi/5}$.

We compared the proposed method with the diagonal loading sample matrix inversion (LSMI) method, middle sub-array based (MSB) beamformer [9] method, orthogonal projection (OP) approach, eigenspace-based (ESB) beamformer method, optimal beamformer method, Li's method [29], and the reconstruction method by Gu [25]. For the proposed beamformer, the RCB uncertainty extent was set as $\epsilon = 2$. For Gu's beamformer (introduced in [25]), Li's method [29], and our proposed beamformer, the desired signal sector was set as $\Theta = [-5^\circ, 5^\circ]$, while the complement sector for the interferences was $\bar{\Theta} = [-90^\circ, -5^\circ) \cup (5^\circ, 90^\circ]$.

The output SINR curves versus the SNR were investigated (with the number of snapshots $K = 500$), as shown in Figure 3. The results clearly show the superiority of the proposed method, which outperformed the others at all SNRs. It is worth noting that the Gu's method is parallel to Optimal SINR method at all SNRs. This is because the deviation between the assumed and real SV structure is determined by the DOA and, therefore, once the DOA distribution of the interferences is set, the deviation between the Gu's method and the optimal is stable at all SNRs. Specifically, when there is no array error, the output SINR of Gu's method can be very close to the optimal. Similarly, Li's method is also parallel to the optimal, since the same INCM reconstruction process is involved. While Gu's method is superior to Li's method, this is due to the array aperture loss in Li's method degrading its performance.

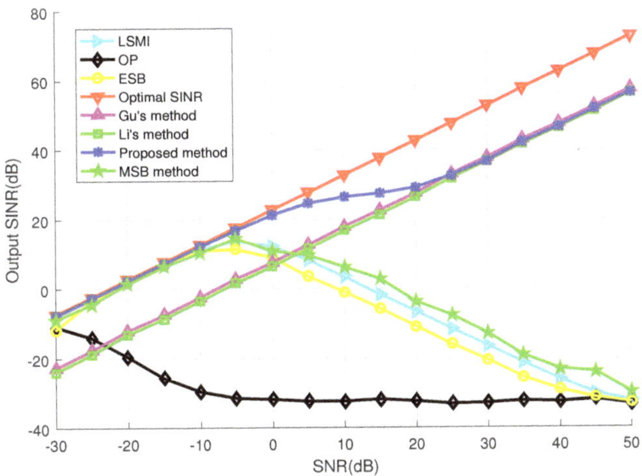

Figure 3. Output signal-to-interference-plus-noise ratio (SINR) versus input signal-to-noise ratio (SNR).

The output SINR of our method is close to the optimal result at low SNRs and outperformed all of the other methods at high SNRs. At low SNRs, the mismatches of the steering vector bring significant influence to the output SINR. The proposed method is able to attain the optimal due to the SV correcting process. It should be noted that, at high SNRs, the performance improvement of the proposed method over other methods decreases, but it still enjoyed the best performance. The proposed method gradually converges to Gu's method at high SNRs. Because, at high input SNRs, the performance increase that arises from the SV correction process gradually decreases, as the higher input SNR is the more important factor for improving the output SINR, rather than the correcting process.

In Figure 4, the gap between the optimal SINR and beamformers are depicted in curves. The deviations from the optimal SINR versus the input SNR can be clearly observed. It can be observed that the proposed method showed similar performance at high SNRs to Gu's method, and it achieved about 18 dB higher at low input SNRs. In terms of Gu's method, its performance was stable, retaining a deviation of about 17 dB from optimal performance. The proposed method achieved fast convergence, while the other beamformers showed slow convergence.

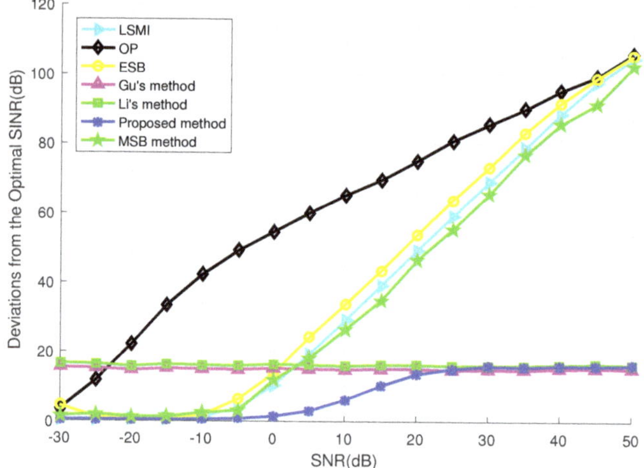

Figure 4. Deviations from the optimal SINR versus input SNR.

With the SNR set at 0 dB, the output SINR versus the number of snapshots is plotted in Figure 5. The depicted curves illustrate the superiority of the proposed method; it was very close to optimal. It can be observed that the proposed method was not sensitive to the number of snapshots and it showed almost the same convergence rate as the optimal beamformer, with the performance improving slightly with an increase of the number of snapshots.

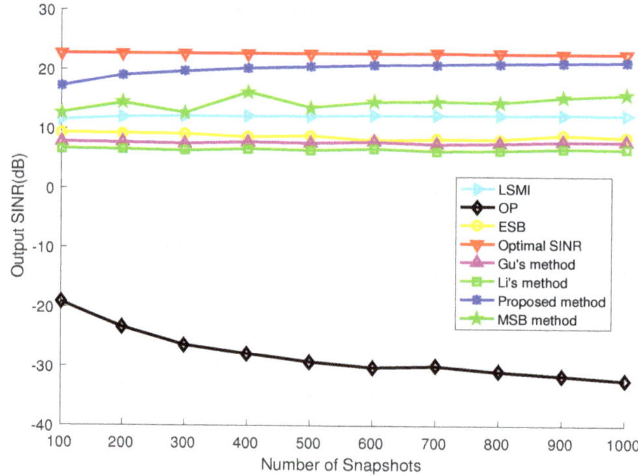

Figure 5. Output SINR versus number of snapshots; SNR = 0 dB, INR = 20 dB.

In the last example, we further investigated the influence of the array aperture on the performance of beamformers. We explored the output SINR curves of different methods with the array length varying from 16 to 35 and the input SNR fixed at 0 dB. Figure 6 clearly shows that, as the array aperture gets larger, the output SINR of the proposed method gets better accordingly. It is also noted that the output SINR of Gu's method as well as the Li's method gets slightly higher with the larger array aperture. The array length reflects the array sampling ability of the signals in the spatial domain. When the array aperture gets larger, the spatial solution of the array gets better and it can more effectively form nulls at the directions that correspond to the interferences. On the other hand, in terms of some other methods, as the solution of the array gets more precise, the mismatches of steering vector become increasingly obvious and degrade the performance more. Therefore, the performance of some methods are slightly getting worse.

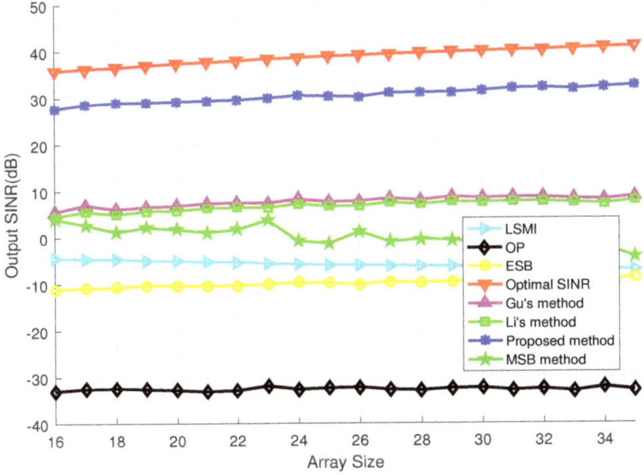

Figure 6. Output SINR versus array aperture; SNR = 0 dB, INR = 20 dB.

5. Conclusions

This paper introduced a new robust adaptive beamforming method, which is robust to the sensor position, gain-phase, and mutual coupling errors. In the proposed method, the mutual coupling effect is mitigated while using the sub-array technique, where the DINCM is reconstructed by combining the corrected SVs and maximum eigenvalue of the DSCM. Moreover, the desired signal SV is obtained using the matched eigenvector. The proposed method is capable of simultaneously dealing with multiple types of array errors. Our simulation results validated the superiority of the proposed method over existing methods in the presence of multiple types of array errors.

Author Contributions: Conceptualization, Z.X. and C.F.; methodology, Z.X.; software, Z.X.; validation, Z.X., C.F. and J.Z.; writing–original draft preparation, Z.X., C.F. and J.Z.; writing–review and editing, C.F., J.Z. and X.H.; supervision, X.H. All authors have read and agree to the published version of the manuscript.

Funding: This research received no external funding.

Conflicts of Interest: The authors declare no conflict of interest.

References

1. Van Trees, H.L. *Optimum Array Processing*; Wiley: New York, NY, USA, 2002.
2. Vorobyov, S.A. Principles of minimum variance robust adaptive beamforming design. *Signal Process.* **2013**, *93*, 3264–3277. [CrossRef]
3. Somasundaram, S.D.; Parsons, N.H. Evaluation of Robust Capon Beamforming for Passive Sonar. *IEEE J. Ocean. Eng.* **2011**, *36*, 686–695. [CrossRef]

4. Somasundaram, S.D. Wideband Robust Capon Beamforming for Passive Sonar. *IEEE J. Ocean. Eng.* **2013**, *38*, 308–322. [CrossRef]
5. Vorobyov, S.A.; Gershman, A.B.; Luo, Z.Q. Robust adaptive beamforming using worst-case performance optimization: A solution to the signal mismatch problem. *IEEE Trans. Signal Process.* **2003**, *51*, 313–324. [CrossRef]
6. Cao, S.; Ye, Z.; Xu, D.; Xu, X. A Hadamard Product Based Method for DOA Estimation and Gain-Phase Error Calibration. *IEEE Trans. Aerosp. Electron. Syst.* **2013**, *49*, 1224–1233.
7. Li, Y.; Er, M. Theoretical analyses of gain and phase error calibration with optimal implementation for linear equispaced array. *IEEE Trans. Signal Process.* **2006**, *54*, 712–723.
8. Rubio, J.; Izquierdo, J.F.; Corcoles, J. Mutual Coupling Compensation Matrices for Transmitting and Receiving Arrays. *IEEE Trans. Antennas Propag.* **2015**, *63*, 839–843. [CrossRef]
9. Ye, Z.; Liu, C. Non-sensitive adaptive beamforming against mutual coupling. *IET Signal Process.* **2009**, *3*, 1–6.
10. Carlson, B.D. Covariance matrix estimation errors and diagonal loading in adaptive arrays. *IEEE Trans. Aerosp. Electron. Syst.* **1988**, *24*, 397–401. [CrossRef]
11. Li, J.; Stoica, P.; Wang, Z. On robust Capon beamforming and diagonal loading. *IEEE Trans. Signal Process.* **2003**, *51*, 1702–1715.
12. Chang, L.; Yeh, C.C. Performance of DMI and eigenspace-based beamformers. *IEEE Trans. Antennas Propag.* **1992**, *40*, 1336–1347. [CrossRef]
13. Feldman, D.D.; Griffiths, L.J. A projection approach for robust adaptive beamforming. *IEEE Trans. Signal Process.* **1994**, *42*, P.867–876. [CrossRef]
14. Feldman, D.D. An analysis of the projection method for robust adaptive beamforming. *IEEE Trans. Antennas Propag.* **1996**, *44*, 1023–1030. [CrossRef]
15. Yu, Z.L.; Gu, Z.; Zhou, J.; Li, Y.; Ser, W.; Er, M.H. A Robust Adaptive Beamformer Based on Worst-Case Semi-Definite Programming. *IEEE Trans. Signal Process.* **2010**, *58*, 5914–5919. [CrossRef]
16. Gershman, A.B.; Sidiropoulos, N.D.; Shahbazpanahi, S.; Bengtsson, M.; Ottersten, B. Convex Optimization-Based Beamforming. *IEEE Signal Process. Mag.* **2010**, *27*, 62–75.
17. Hassanien, A.; Vorobyov, S.A.; Wong, K.M. Robust Adaptive Beamforming Using Sequential Quadratic Programming: An Iterative Solution to the Mismatch Problem. *IEEE Signal Process. Lett.* **2008**, *15*, 733–736. [CrossRef]
18. Vorobyov, S.A.; Gershman, A.B.; Luo, Z.-Q.; Ma, N. Adaptive beamforming with joint robustness against mismatched signal steering vector and interference nonstationarity. *IEEE Signal Process. Lett.* **2004**, *11*, 108–111. [CrossRef]
19. Stoica, P.; Wang, Z.; Li, J. Robust Capon beamforming. *IEEE Signal Process. Lett.* **2003**, *10*, 172–175.
20. Li, J.; Stoica, P.; Wang, Z. Doubly constrained robust Capon beamformer. *IEEE Trans. Signal Process.* **2004**, *52*, 2407–2423.
21. Liao, B.; Guo, C.; Huang, L.; Li, Q.; So, H.C. Robust Adaptive Beamforming With Precise Main Beam Control. *IEEE Trans. Aerosp. Electron. Syst.* **2017**, *53*, 345–356. [CrossRef]
22. Beck, A.; Eldar, Y.C. Doubly Constrained Robust Capon Beamformer With Ellipsoidal Uncertainty Sets. *IEEE Trans. Signal Process.* **2007**, *55*, 753–758. [CrossRef]
23. Zhuang, J.; Huang, P. Robust Adaptive Array Beamforming With Subspace Steering Vector Uncertainties. *IEEE Signal Process. Lett.* **2012**, *19*, 785–788. [CrossRef]
24. Vorobyov, S.A.; Chen, H.; Gershman, A.B. On the Relationship Between Robust Minimum Variance Beamformers With Probabilistic and Worst-Case Distortionless Response Constraints. *IEEE Trans. Signal Process.* **2008**, *56*, 5719–5724. [CrossRef]
25. Gu, Y.; Leshem, A. Robust Adaptive Beamforming Based on Interference Covariance Matrix Reconstruction and Steering Vector Estimation. *IEEE Trans. Signal Process.* **2012**, *60*, 3881–3885.
26. Gu, Y.; Goodman, N.A.; Hong, S.; Li, Y. Robust adaptive beamforming based on interference covariance matrix sparse reconstruction. *Signal Process.* **2014**, *96*, 375–381. [CrossRef]
27. Huang, L.; Zhang, J.; Xu, X.; Ye, Z. Robust Adaptive Beamforming With a Novel Interference-Plus-Noise Covariance Matrix Reconstruction Method. *IEEE Trans. Signal Process.* **2015**, *63*, 1643–1650. [CrossRef]
28. Zheng, Z.; Yang, T.; Wang, W.Q.; So, H.C. Robust adaptive beamforming via simplified interference power estimation. *IEEE Trans. Aerosp. Electron. Syst.* **2019**, *55*, 3139–3152. [CrossRef]

29. Li, Z.; Zhang, Y.; Ge, Q.; Guo, Y. Middle Subarray Interference Covariance Matrix Reconstruction Approach for Robust Adaptive Beamforming With Mutual Coupling. *IEEE Commun. Lett.* **2019**, *23*, 664–667. [CrossRef]
30. Capon, J. High-resolution frequency-wavenumber spectrum analysis. *Proc. IEEE* **1969**, *57*, 1408–1418. [CrossRef]
31. Harmanci, K.; Tabrikian, J.; Krolik, J.L. Relationships between adaptive minimum variance beamforming and optimal source localization. *IEEE Trans. Signal Process.* **2000**, *48*, 1–12. [CrossRef]
32. Chen, F.; Shen, F.; Song, J. Robust adaptive beamforming using low-complexity correlation coefficient calculation algorithms. *Electron. Lett.* **2015**, *51*, 443–445.

MDPI
St. Alban-Anlage 66
4052 Basel
Switzerland
Tel. +41 61 683 77 34
Fax +41 61 302 89 18
www.mdpi.com

Journal of Marine Science and Engineering Editorial Office
E-mail: jmse@mdpi.com
www.mdpi.com/journal/jmse

www.ingramcontent.com/pod-product-compliance
Lightning Source LLC
LaVergne TN
LVHW070653170726
843515LV00004B/398

* 9 7 8 3 0 3 6 5 5 4 9 7 6 *